Surface Engineering for Biomedical Applications (SEBA 2020)

IOP Conference Series: Materials Science and Engineering
Volume 1056

Pleven, Bulgaria
11 November 2020

ISBN: 978-1-7138-3338-3
ISSN: 1757-8981

Printed from e-media with permission by:

Curran Associates, Inc.
57 Morehouse Lane
Red Hook, NY 12571

Some format issues inherent in the e-media version may also appear in this print version.

Additional copies of this publication are available from:

Curran Associates, Inc.
57 Morehouse Lane
Red Hook, NY 12571 USA
Phone: 845-758-0400
Fax: 845-758-2633
Email: curran@proceedings.com
Web: www.proceedings.com

TABLE OF CONTENTS

IOP Conf. Series: Materials Science and Engineering 1056 (2021) 011001 doi:10.1088/1757-899X/1056/1/011001

Preface
IOP Conference Series: Materials Science and Engineering

1ST ON-LINE WORKSHOP

SURFACE ENGINEERING FOR BIOMEDICAL APPLICATIONS

ENGINEERING WITH A MEDICAL PERSPECTIVE: FUTURE BEGINS TODAY

On **11.11.2020** the first online seminar **"SURFACE ENGINEERING FOR BIOMEDICAL APPLICATIONS 2020"** – **SEBA2020** was held. The complicated world situation forced the first edition of the scientific event to take place in the form of a virtual forum. New materials and techniques for application in biomedicine and the modern engineering technologies were discussed for the first time at the virtual scientific event by scientists from Bulgaria, Belgium, Slovenia, Spain, Portugal and India. Organizers of the scientific seminar **SEBA 2020** (Engineering of surfaces for application in biomedicine) were the Institute of Electronics - BAS and the Medical University of Pleven.

INSTITUTE OF ELECTRONICS
BULGARIAN ACADEMY OF SCIENCES
72 Tsarigradsko Chaussee, 1784 Sofia, Bulgaria
Phone: +359 2 875 0077, Fax: +359 2 975 32 01, http://www.ie-bas.org

The Institute of Electronics carries out research and educational activities in the fields of Physical Electronics, Quantum Electronics and Radiophysics. The research is focused on: studying the processes of generation and control of beams of electrons, ions and photons, and especially on their interaction with matter; development of experimental and industrial equipment for surface modification, thin-film deposition and analysis; welding and melting of metals; low-temperature plasma, and plasma-chemical processes; plasma-assisted formation of thin films and coatings; interaction of laser radiation with matter and thin-film formation; development of laser sources and systems for spectroscopy, metrology, modification and analysis of materials; non-linear optical phenomena in fibers and semiconductors; generation and propagation of electromagnetic waves; non-linear phenomena in a wide MW-frequency range; scattering of electromagnetic waves from homogeneous and inhomogeneous media, signal detection, data acquisition and processing; development of electronic systems for the MW and optical ranges; systems for remote sounding of the atmosphere and sending of the sea and Earth surface.

IOP Conf. Series: Materials Science and Engineering 1056 (2021) 011001 doi:10.1088/1757-899X/1056/1/011001

MEDICAL UNIVERSITY - PLEVEN
1, Sv. Kliment Ohridski Str. 5800 Pleven, Bulgaria
Fax: (++359 64) 801 603, Telephone, Rector's Office: (++359 64) 884 101
http://www.mu-pleven.bg/index.php/bg/

Medical University - Pleven, established in 1974, is one of the most dynamic and advanced higher medical educational institutions in Bulgaria. The University consists of 4 faculties - Faculty of Medicine, Faculty of Pharmacy, Faculty of Public health, Faculty of Health care. Medical College and Department of Language and Specialized Training are administrative units of MU-Pleven.

Today, combining traditions of the past with the present possibilities, the University incorporates educational and therapeutic facilities, contemporary pre-clinical base, University Hospital and clinics in all major medical fields, as well as a large number of specialized clinics and research units with modern diagnostic and therapeutic equipment. These are efficiently used to treat patients, train students, trainee doctors, post-graduates as well as for research work.

Medical University – Pleven is growing and developing very fast, with an emphasis on implementing advanced technologies in education and clinical practice. The medical school has two robotic systems for performing robot-assisted surgery with simulators for training and research. This ranks MU-Pleven among the leading European educational centers in robotics surgery. The medical school is advancing the practice of robotic surgery, performing Live Surgery in 3D format at modern Telecommunication Endoscopic Center. This facility, which houses its own dedicated experimental operating room, is an unique emblem of the MU-Pleven within the country and abroad.

The one-day virtual forum was held on November 11, 2020, under the motto **"Engineering with a medical perspective: the future begins today"**. The presented reports were divided into three sections: Bioengineering and fundamental applications; New materials and techniques for biomedical applications; Application of modern engineering technologies in the medicine. **BigBlueButton** was used as a virtual platform for SEBA 2020:
https://bbb.mu-pleven.bg/b/teo-663-239
The final scientific program of the event included presentations by **7 invited lecturers** from Bulgaria, Portugal, Spain and Slovenia and **22 oral presentations**.
The international program committee has selected some papers for publishing in the IOP Conference Series: Materials Science and Engineering. Every submitted paper is subjected to a rigorous reviewing process. The main criterion is a judgment of the degree to which the submitted paper contributes substantial new research. Reviewers rate the papers submitted by a 100% ranking scale for this purpose and provide a written evaluation of the manuscripts.

List of participants in SEBA 2020

	Invited speakers	**Afiliation**
1	Prof. Amelia Almeida	Technical University of Lisbon, Lisbon, Portugal
2	Prof. Vera Marinova	Institute of Optical Materials and Technologies – Bulgarian Academy of Sciences
3	Prof. Miran Mozetič	Jozef Stefan Institute, Ljubljana, Slovenia
4	Prof. Endzhe Matykina	Universidad Complutense de Madrid, Madrid, Spain

5	Prof. Dr. Slavcho Tomov, MD, PhD, DSc	Medical University of Pleven
6	Prof. Grigor Gorchev, MD, PhD, DSc	Honorary Rector of the Medical University, Pleven
7	Prof. Rui Vilar	Technical University of Lisbon, Lisbon, Portugal
	Oral presentations	
8	Assoc. Prof. Albena Daskalova, PhD	Institute of Electronics, BAS
9	Assoc. Prof. Ekaterina Iordanova, PhD	Institute of Solid State Physics, BAS
10	Shiva Udachan	Rani Channamma University, Karnataka, India
11	Assoc. Prof. Anna Dikovska, PhD	Institute of Electronics, BAS
12	Meriam Ben Abdeljawad	Laboratory of Polymeric and Composite Materials (SMPC), Center of Innovation and Research in Materials and Polymers (CIRMAP), University of Mons, Belgium
13	Xavier Carette	Laboratory of Polymeric and Composite Materials (LPCM), University of Mons, Belgium
14	Angela Carvalho1,2	1i3S - Instituto de Investigação e Inovação em Saúde, U. Porto, Portugal; 2INEB - Instituto de Engenharia Biomédica, U. Porto, Portugal
15	Assoc. Prof. Maria Nikolova, PhD	Dept. of Material Science and Technology, University of Ruse "A. Kanchev", Ruse, Bulgaria
16	Svetoslava Stoycheva	Department of Biochemistry, Medical University-Pleven, Bulgaria
17	Chief Assist. Prof. Stefan Valkov, PhD	Institute of Electronics, BAS
18	Ralitsa Angelova, PhD	Institute of Electronics, BAS
19	Hristina Galeva	Faculty of Dental medicine, Medical University, Sofia, Bulgaria
20	Assoc. Prof. Nikolai Donkov, PhD	Institute of Electronics, BAS
21	Iva Taneva	Faculty of Dental medicine, Medical University, Sofia, Bulgaria
22	Chief Assist. Prof. Stoyan Tsvetkov, PhD	Institute of Electronics, BAS
23	Assoc. Prof. Nataliya Chilingirova, MD, PhD	Science and Research Institute, Medical University – Pleven, Bulgaria
24	Prof. Ivan Nedkov, DSc	Institute of Electronics, BAS
25	Vladislav Nankov	University Scientific Research Laboratory, Medical University – Pleven, Bulgaria
26	Prof. Dobromir Dimitrov, MD, PhD	Medical University – Pleven, Bulgaria
27	Prof. Katia Kovacheva, MD, PhD	Medical University – Pleven, Bulgaria

IOP Conf. Series: Materials Science and Engineering 1056 (2021) 011001 doi:10.1088/1757-899X/1056/1/011001

ORGANIZING COMMITTEE
Presidents
Assoc. Prof. Galya Stavreva, MD, PhD
Vice-Rector of Science and Research, Medical University of Pleven
Assoc. Prof. Albena Daskalova, PhD
Researcher in laboratory „Micro and Nano photonics", BAS

Members
Assoc. Prof. Teodora Valova, PhD
Assoc. Prof. Nadya Veleva, PhD
Chief Assist. Prof. Maria Ormanova, PhD
Assist. Prof. Stanislava Rabadzhiyska, PhD
Liliya Angelova, PhD student
Vladislav Nankov

SCIENTIFIC COMMITTEE
Presidents:
Prof. Dr. Slavcho Tomov, MD, PhD, DSc
Rector, Medical University of Pleven
Prof. Petar Petrov, PhD, DSc
Director, Academician Emil Djakov Institute of Electronics, Bulgarian Academy of Sciences

Members
Prof. Grigor Gorchev, MD, PhD, DSc
Prof. Silviya Yankulovska, MD, PhD, DSc, MAS
Prof. Ivan Petrov, University of Illinois, Urbana, IL, USA
Prof. Rui Vilar, IST, Lisbon, Portugal

INVITED SPEAKERS
Prof. Amelia Almeida, Technical University of Lisbon, Lisbon, Portugal
Prof. Rui Vilar, IST, Lisbon, Portugal
Prof. Vera Marinova, Institute of Optical Materials and Technologies – Bulgarian Academy of Sciences
Prof. Miran Mozetič, Jozef Stefan Institute, Ljubljana, Slovenia
Prof. Endzhe Matykina, Universidad Complutense de Madrid, Madrid, Spain
Prof. Dr. Slavcho Tomov, MD, PhD, DSc, Medical University of Pleven
Prof. Grigor Gorchev, MD, PhD, DSc, Honorary Rector of the Medical University, Pleven

On behalf of the organizing committee I would like to express sincere appreciation to authors, reviewers, participants and to all the organizers team for their enthusiasm and support which granted success to **SEBA2020**.
Chair of the Organizing committee
Assoc. Prof. Albena Daskalova

Peer review declaration

All papers published in this volume of *IOP Conference Series: Materials Science and Engineering* have been peer reviewed through processes administered by the Editors. Reviews were conducted by expert referees to the professional and scientific standards expected of a proceedings journal published by IOP Publishing.

Type of peer review: Single-blind

- **Conference submission management system:**

The papers were submitted to the conference secretariat and then distributed to reviewers by the editors. The reviewers sent back their reviews to the editor who then finalized the editing process.

- **Number of submissions received: 16**
- **Number of submissions sent for review: 16**
- **Number of submissions accepted: 15**
- **Acceptance Rate (Number of Submissions Accepted / Number of Submissions**

- **Received X 100): 93,75%**
- **Average number of reviews per paper: 1**
- **Total number of reviewers involved: 10**
- **Any additional info on review process (ie plagiarism check system):** The paper was checked if it is relevant to the topic of the webinar and contains new research result(s), if it is clearly written and structured; figures and tables are informative and correctly organized; references are correct and contain new literature sources; the formatting according to IOP standard was applied. After revision the papers were submitted to the editor.

Contact person for queries: Assoc. Prof. Albena Daskalova, PhD
Institute of Electronics, Bulgarian Academy of Sciences, Sofia, Bulgaria
e-mail: albdaskalova@gmail.com

INFLUENCE OF FEMTOSECOND LASER PROCESSING PARAMETERS ON THE SURFACE MORPHOLOGY AND WETTABILITY PROPERTIES OF POLYLACTIC ACID

Albena Daskalova[1*], Liliya Angelova[1], Xavier Carette[2], Rosica Mincheva[2], Jean-Marie Raquez[2], Anton Trifonov[3], Ivan Buchvarov[3]
[1]Institute of Electronics, Bulgarian Academy of Sciences, Sofia, Bulgaria
[2]Laboratory of Polymeric and Composite Materials (LPCM), University of Mons, Place du Parc, 23, B-7000, Mons, Belgium
[3]Faculty of Physics, St. Kliment Ohridski University of Sofia, Sofia, Bulgaria
*e-mail: albdaskalova@gmail.com

Abstract. Polylactic acid (PLA) is a very attractive material for applications in the field of regenerative medicine and tissue engineering, as it is a biocompatible and biodegradable synthetic polymer, whose morphological properties can be finely tuned via laser processing. Ultra-short pulse laser treatment is a non-invasive method for optimizing the surface properties of engineered biomatrices. The method offers controlled porosity of the treated biomaterial tailored specifically for implantation needs. In this research, we investigated the interaction of femtosecond laser radiation with PLA stacks while varying the output laser parameters in order to estimate their influence on the morphology of the PLA samples. The induced microstructural features were characterized by thickness measurements and SEM, EDX and FTIR analyses. In order to evaluate the hydrophilicity of the treated surface, water contact angle (WCA), measurements were also performed. Topography modification of the PLA substrates could essentially improve this material's bioactivity properties, which, after proper optimization of the laser parameters, could make its biomedical applications even more successful.

1. Introduction

Regenerative medicine and, in particular, tissue engineering represent the basis of today's and tomorrow's medicine [1,2]. In this field, the polylactic acid (PLA) is a very attractive newly emerging material with the potential to become a basic one, as it is an easily produced biocompatible and biodegradable polymer, whose morphological qualities can be further improved [3,4]. In their exhaustive review, Khoddami et al. [5] explained in detail all the advantages of PLA as a promising material for bone tissue engineering – PLA interacts well very easily with the host bone cells when it is implanted [6,7]. This fact makes it an ideal candidate for contact surface material between the natural bone and the external implant and for a bone scaffold substrate, as the ideal engineered matrix should be composed of biomaterials that closely mimic the structure and qualities of natural bone extracellular matrix [8,9]. The mechanical properties of PLA are similar to those of bone tissues and, what is more, it is amenable to processing and degrades into natural metabolites, which makes it nontoxic for the organism [10]. Ultra-short pulse laser treatment is a non-invasive method that perfectly fits into the idea of optimizing the surface properties of cell-engineered matrices, as it can offer strictly controlled porosity of the treated biomaterial tailored specifically for implantation, without adding unnatural physical or chemical components to the scaffold created. This topography-reforming biocompatible technique successfully overcomes the limitations associated with the application of other physical and chemical methods since the side effects caused by the interaction of fs laser pulses with biological tissues are minimized. Moreover, ultra-short pulse laser treatment provides a porous cell scaffold, with precisely varied dimensions of the created interconnected pores ($1 \div 500$ μm), providing diffusion of oxygen and nutrients, as well as cell viability, adhesion, movement and tissue ingrowth into the engineered ECM [11,12].

The main goal of this study is to acquire initial knowledge of the porosity and morphology of PLA samples after ultra-fast laser modification in view of future design and preparation of porous PLA-based cell matrices, which could form the basis for creating effective bio-interfaces between the recipient and the foreign implant. For this purpose, fs modified microstructured polymer scaffolds were investigated by SEM, EDX and FTIR. Wettability measurements of treated and non-treated surfaces were performed. The thicknesses of the samples before and after the ultra-fast laser modification were also compared.

2. Material and methods

2.1. Sample preparation
The PLA is supplied by Nature Works (PLA 4060D) and is amorphous with a *Tg* of around 60 °C. PLA pellets were prepared via heat compression molding using a Carver 4122 12-12H manual heated press. The pellets were first vacuum-dried overnight at 60 °C, then inserted in the mold (a 0.5-mm thick square). The following compression procedure was used: at 180 °C, 3 min of contact, several steps of degassing, and 2 min at 12 bar. The material was cut into 4×4 cm^2 squares.

2.2. Laser treatment
The femtosecond ($\tau = 150$ fs) laser modification of the prepared PLA samples was performed by a Quantronix-Integra-C Ti:sapphire laser system oscillating at a central wavelength of $\lambda = 800$ nm and 0.5-kHz repetition rate. The applied energy and number of laser pulses were varied as follows: $F = 0.42$ J/cm^2 and 0.34 J/cm^2; $N = 2$, 10, and 30. The experiments were performed in air and the setup was controlled by LabView software. The sample was positioned normally to the laser beam on a high-precision XY translation stage and processed by scanning the laser beam over the material's surface at precisely defined separation intervals in order to optimize the PLA surface texturing.

2.3. Thickness measurement
The thickness of the tested samples was measured before and after the laser treatment to determine the penetration depth of the fs pulses applied. For this purpose, a coating meter (coating thickness gauge VA 8042) was used, with every value quoted representing the average of 10 separate measurements.

2.4. SEM-EDX analysis
The surface morphology and elemental composition before and after the fs laser treatment were analyzed by scanning electron microscopy (SEM) with energy dispersive X-ray analysis (EDX) (TESCAN/LYRA/XMU, respectively). The samples were gold-sputtered and several images were taken for every modification investigated. The corresponding elemental composition [wt. %] was also defined.

2.5. WCA evaluation
Water contact angle evaluation of the control and fs-modified PLA was performed at room temperature by a homemade installation in air – 1 µl dH_2O drop was applied on the samples surface and images were taken every 0.5 s for 9 s. The contact angles were measured using the ImageJ software. All contact angle values were averaged according to the patterning type.

2.6. FTIR analysis
A Fourier-transform infrared (FTIR) spectrophotometer (IR Affinity-1, Shimadzu, Kyoto, Japan), working in the range from 4500 cm^{-1} to 500 cm^{-1} was used to obtain the IR transmittance spectra [%] of the prepared probes to evaluate the chemical bonds alterations after laser treatment relative to the control sample surface.

3. Results and discussion
The surface morphology of the PLA sheets after femtosecond irradiation at different fluences and multiple number of laser pulses is presented below. Figure 1 shows representative SEM images and the corresponding EDX elemental composition [wt. %] of a control PLA sheet and PLA sheets patterned by $N = 2$, 10, and 30 fs pulses and laser fluence $F = 0.34$ J/cm^2 and 0.42 J/cm^2.

(a)

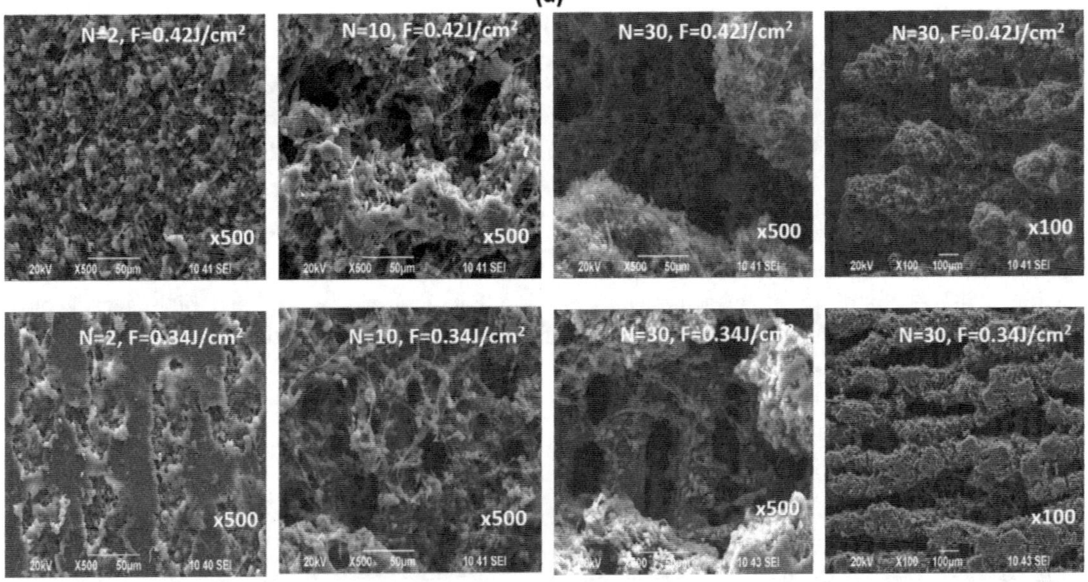

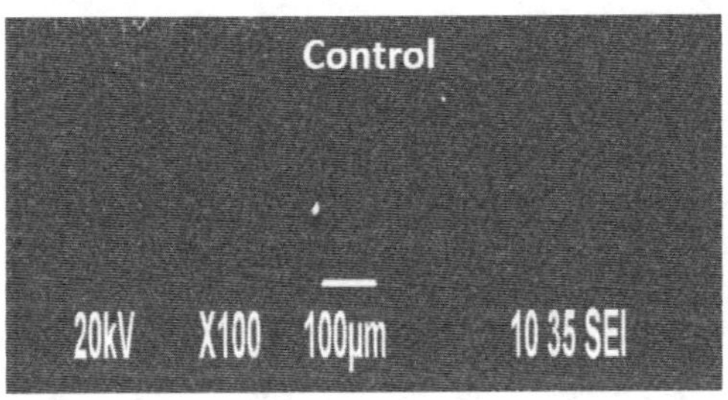

Sample	C [wt. %]	O [wt. %]
PLA control	74.27	25.73
PLA N=2, F=0.42J/cm^2	61.51	38.49
PLA N=10, F=0.42J/cm^2	60.52	39.48
PLA N=30, F=0.42J/cm2	58.82	41.18
PLA N=2, F=0.34J/cm^2	60.42	39.58
PLA N=10, F=0.34J/cm^2	59.09	40.91
PLA N=30, F=0.34J/cm^2	58.32	41.68

(b) **(c)**

Figure 1. SEM images of PLA matrices (a) patterned with $N = 2$, 10, 30 and $F = 0.42$ J/cm^2 (first row) and $F = 0.34$ J/cm^2 (second row); (b) control PLA surface; (c) EDX elemental composition [wt. %] of the corresponding samples.

Comparing panels (a) and (b) of Figure 1 one can easily see that the laser-processed PLA samples exhibit a microporous granular morphology for all parameters of fs modification.

After processing by $N = 2$ at $F = 0.34$ J/cm^2, adhesion of raised spatters around the sides of interaction zone is seen. Increasing F from 0.34 J/cm^2 to 0.42 J/cm^2 results in smoother zones, while increasing the number of applied laser pulses from 10 to 30 leads to the formation of through holes. Using either value of F and the highest number of laser pulses initiates the creation of self-formed complex frameworks. The microstructures generated when using a large number of pulses exhibit a periodicity. This result could have positive effect on cells' behavior, as the modification affects the scaffold morphology in depth and thus provides a greater porosity in the volume. The slight variation in the [C] and [O] [wt. %] with changing N and F of the fs pulses could be explained with the breakage of weak chemical bonds, which does not affect the elemental composition of the samples, as no elements uncommon for PLA are observed in the EDX data, Figure 1(c). In this way, the ultra-fast laser processing provides precise control of the shape and the size of the hierarchically connected pores without affecting the chemical composition of the biomaterial in order to further resemble the natural cellular environment. The results from the WCA measurements show that the material changes its wettability properties. The WCA values of the control sample was measured to be $\theta = 74°$.

Increasing the number of laser pulses to $N = 2$ for both laser fluences applied reduces the average contact angle. When using a large number of pulses to modify the material, the initial WCA values are the highest in the first second compared to the WCA obtained from patterns acquired with fewer N; this corresponds to a morphology having a micro-scale hierarchical structure, Figure 2 (a and b). The microstructures on the processed PLA surface obtained under irradiation with multiple pulses have a decreased thickness (Table 1); thus, when the water droplet bounds to the surface, it undergoes heterogeneous wetting. Due to the increased surface roughness, vapor pockets may be trapped underneath the liquid. This heterogeneous wetting is usually described by the Cassie-Baxter (CB) model, which is proved by the values obtained for WCA for $N = 30$.

In this way, the functionality of the cells implant may be further improved, as control over cells adhesion and movement could be achieved. Apart of that, hydrophilic grooved biomaterials have improved surface qualities, making them a preferable choice for covering traditional long-lasting bone implants.

(a)

Time [s]	PLA N=2, F=0.42J/cm²	PLA N=10, F=0.42J/cm²	PLA N=30, F=0.42J/cm²	PLA N=2, F=0.34J/cm²	PLA N=10, F=0.34J/cm²	PLA N=30, F=0.34J/cm²
0.5	80.1°	136.5°	83.35°	85.8°	83.9°	123.75°
1	75°	95.1°	77°	81.9°	75°	121.55°
3	67.2°	81.65°	76.3°	79.2°	74.35°	119.35°
5	59.65°	73.4°	72°	71.45°	61.8°	111.6°
7	56.05°	60.05°	62.3°	70.35°	61.75°	111.5°
9	53.05°	56.35°	60.95°	69.5°	49.5°	80.7°

(b)

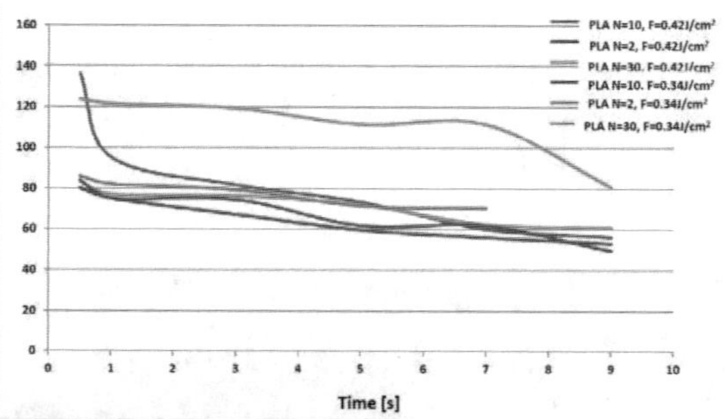

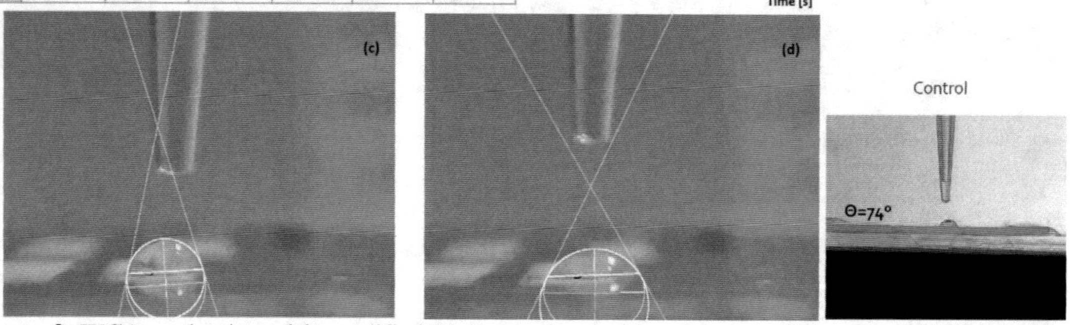

Figure 2. WCA evaluation of fs-modified PLA samples (a, b) and image of the water droplet on the surface of fs-modified PLA, $N = 10$, $F = 0.34$ J/cm^2 after 0.5 s (c) and 5 s (d).

The thickness of the tested samples before and after laser treatment is quoted in Table 1. As can be seen, the thickness of the laser modified surface areas decreases as the number of applied laser pulses is increased. This finding is in agreement with the morphological examinations and is due to a melting of the solid polymer matrix and severe ablation leading to formation of wells and islands with conical-like shape when the surface temperature suddenly increased during the ablation. The partially melted phase is followed by resolidification, which affects the processed region thickness in the case of high laser fluence values ($F = 0.42$ J/cm^2).

PLA	Thicknes s [μm]	PLA	Thickness [μm]
Control	501		
$N = 2$, $F = 0.42$ J/cm^2	491	$N = 2$, $F = 0.34$ J/cm^2	472
$N = 10$, $F = 0.42$ J/cm^2	486	$N = 10$, $F = 0.34$ J/cm^2	459
$N = 30$, $F = 0.42$ J/cm^2	479	$N = 30$, $F = 0.34$ J/cm^2	448

Table 1. PLA samples thickness measurement before and after fs modification with different fluence and number of pulses applied.

The FTIR spectroscopy results (Figure 3) after fs laser treatment compared with the untreated PLA [13] confirm the lack of changes in the chemical composition of the samples. The shapes of the spectra from the processed PLA areas for both fluences and different number of applied laser pulses do not show discrepancies, the only difference being the intensity of the peaks. The regions of interest are visible in all spectra of the fs-modified PLA. The maximum at 3600–4000 cm^{-1} characteristic for the PLA corresponds to O–H bond stretching. The peaks detected at 1750 cm^{-1} and 1180 cm^{-1} correspond to the C=O stretching and the C–O–C stretching of PLA. They are clearly visible in all PLA spectra. The band at 1780 cm^{-1} is attributed to stretching of the C=O bond. The carbonyl peak at 1650 cm^{-1} is also well defined.

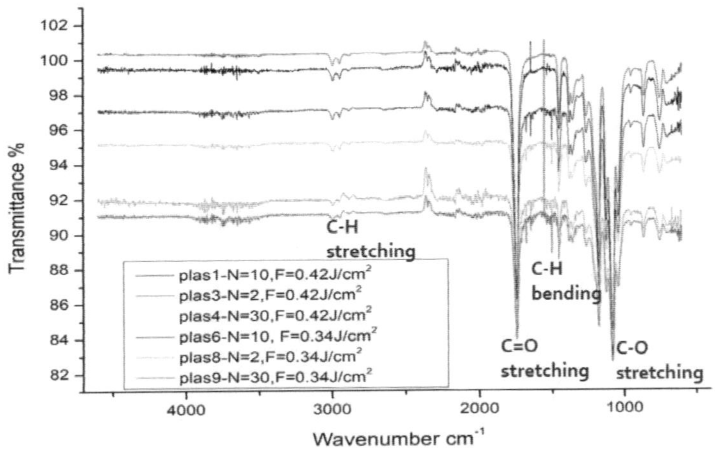

Figure 3. FTIR spectra of PLA matrices after fs laser irradiation.

4. Conclusion

Formation of a smooth surface is observed on the PLA surface when the least number of irradiating pulses is used. The generation of micro-scale hierarchical structures due to the emission of gas-phase products when the number of laser pulses is drastically increased lead to configurations of self-formed periodical structures. The results obtained confirm that fs laser irradiation is a promising surface processing technique for future design of porous PLA-based cell matrices, which could be the basis for creating effective scaffolds. The modification of the PLA substrates topography could essentially improve the bioactivity properties of this material, which, after a proper optimization of the laser parameters, could make its biomedical applications even more successful.

Acknowledgments

The authors acknowledge the financial support under projects KP-06-RILA/8 "Ultra-short laser based functionalization of bioceramics and composites" (2018-2020), KP-06-H 38/5 (2019-2022), "Functionalization of 3D printed fibrous scaffolds via femtosecond laser patterning".

References

[1] Cortesini R 2005 *Transplant Immunology* vol 15 pp. 81–89
[2] Puppi D Chiellini F Piras A M and Chiellini E 2010 *Prog. Polym. Sci.* vol. 35 pp 403-440
[3] Tehrani A H Singh S and Oloyede A 2014 *In: 23rd Annual Conference of the Australasian Society of Biomaterials and Tissue Engineering,* Mantra Resort Lorne, VIC

[4] Chahal S Hussain F S J and Yusoff M M 2013 *Procedia Eng.* vol 53 pp 683–688
[5] Khoddami et al. 2019 *Jornal of Textiles and Polymers* vol 7
[6] Khoddami A Avinc O and Ghahremanzadeh F 2011 *Prog. Org. Coat.* vol 72 pp 299-304
[7] Fattahi F Izadan H and Khoddami A *2011 In: 4th International Color and Coatings Congress (ICCC 2011)* Tehran, Iran, pp 22-24
[8] Sharma J Lizu M Stewart M Zygula K Lu Y Chauhan R Yan X et al. 2015 *Polymers* vol 7 pp 186-219
[9] Kumbar S G Prasad S Nukavarapu James R Nair L S and Laurencin C T 2008 *Biomaterials* vol 29 pp 4100-4107
[10] http://www.natureworksllc.com.
[11] Zeltinger J Sherwood J Graham D Müeller R and Griffith L 2001 Tissue Eng. vol 7 pp 557-572
[12] Turnbull G Clarke J Picard F Riches P Jia L Han F Li B and Shu W 2017 Bioactive Materials vol 3 pp 278-314
[13] Mofokeng J P Luyt A S Ta´bi T and Kova´cs J 2012 *Journal of Thermoplastic Composite Materials* vol 25 pp 927–948

Ultra-short laser pulse modification of chitosan/silver nanoparticles (AgNPs) thin films for potential antimicrobial applications

E Iordanova[1], G Yankov[1], A Daskalova[2], A Dikovska[2], L Angelova[2], D Aceti[2], E Filipov[2], G Stanev[2], B Calin[3], M Zamfirescu[3]

[1]Institute of Solid State Physics, Bulgarian Academy of Sciences,
72 Tsarigradsko Chaussee blvd., 1784 Sofia, Bulgaria
[2]Institute of Electronics, Bulgarian Academy of Sciences,
72 Tsarigradsko Chaussee blvd., Sofia 1784, Bulgaria
[3]National Institute for Laser, Plasma and Radiation Physics, INFLPR, Bucharest, Romania

Corresponding author: eiordanova@issp.bas.bg

Abstract. The last several years witnessed increasingly rapid advances in applying biopolymers in tissue engineering for the purposes of regenerative medicine. The growing demand for preparing materials with desired physical, biological and mechanical properties requires active investigations in the field of tissue engineering. Among the wide variety of biopolymers, chitosan proved to be an outstanding material due to its properties, such as biocompatibility, biodegradability and a wide range of available fabrication technologies. The present work is a case study of an extensive research on the functionalization of thin biopolymer films via laser patterning. The aim of the current study is to investigate the optical properties of biopolymer films on the example of chitosan and chitosan with silver nanoparticles additives. As laser sources are used a nano- and a femto-second laser system emitting the wavelengths of 355 nm and 800 nm, respectively. The compositions produced are investigated by spectral analyses using spectrometers and an optical microscope. Furthermore, the morphology of the samples is monitored by SEM analyses. The results obtained demonstrate the potential of the method employed for obtaining diverse porous modifications depending on the laser parameters. Adding silver nanoparticles will drastically increase the thin chitosan films' antimicrobial properties, thus enhancing the biocompatibility properties of the 2D matrices created.

1. Introduction

In the search for alternatives to the conventional treatment strategies for the repair or replacement of missing or malfunctioning human tissues and organs, promising solutions have been explored through tissue engineering approaches. Tissue engineering offers solutions for improved functionality and biocompatibility of material implants inside the biological environment.

In the last few decades, a range of biomaterials, such as different polymers and bioceramics, have gradually established themselves as alternatives to standard materials (metals and metal sleepers) for permanent replacement and temporary stabilization of orthopedic implants, providing better adaptability, less side effects and faster recovery. At present, the efforts are mainly focused on the development of synthesized polymeric materials that are as close as possible to the chemistry, structure and function of the natural tissue. The choice of biomaterial is one of the most important steps in the process of manufacturing implants for regenerative medicine. Among the variety of biopolymers chitosan has been shown to be an excellent material due to its distinctive properties, such as biocompatibility, biodegradability and a wide range of fabrication technologies [1-3]. The chitosan can be processed into films and porous structures for applications in regenerative medicine, drug delivery, wound bandaging, blood vessels, and nerve conduits [4-7]. Furthermore, the chitosan is being considered as a potential candidate for bio-dental applications due to its antimicrobial properties and ability to blend with other materials [8]. Studies have been carried out with the use of biopolymers, such as chitosan, gelatin and hyaluronic acid, as precursors for tissue regeneration in bone and cartilage engineering. It is known that chitosan itself has antimicrobial activity due to its cationic properties causing a membrane-disrupting effect. There are theories suggesting that the increased antimicrobial activity of Ag-incorporated nanoparticles in chitosan arise from the high degree of infiltration of the silver component with a resulting high bactericidal effect. Recently, scientists have begun to synthesize chitosan-based nanocomposites, as they demonstrate various advantages, such as antibacterial activity, biocompatibility, higher stability and better cell response. These properties have increased the interest in chitosan nanocomposites that could provide a novel way of developing new functional materials for the purposes of regenerative medicine.

A further step in improving the properties of a biomaterial can be achieved by modification. Over the years, various physical and chemical patterning approaches have been employed for surface structuring, including photolithography, electron-beam lithography, electrochemical deposition, selective growth of carbon nanotubes and plasma treatment [9, 10]. In recent years, nano- and micro-patterned surfaces have attracted an increasing attention among the scientific communities of physics, chemistry, medicine and biology [11-13]. It has been shown that the surface chemistry and topography of biomaterial substrates not only affect strongly the cell morphology [14], but also influences the cell behavior, such as adhesion, migration, orientation, guidance, differentiation, proliferation, as well as gene expression and protein synthesis [15]. Understanding the bacteria-surface interaction mechanism is essential for the design of biomaterial surfaces with improved properties.

The morphological changes of the surfaces of biofilms and 3D polymeric scaffolds affect the cells growth and adhesion and their distribution in a desired direction. In this respect, lasers provide significant advantages thanks to the precise control of the laser power, e.g., avoiding heating and melting and the consequent blockage of the porous structure, as well as controlling cell attachment, proliferation, and differentiation. Laser modification can be used to shape materials both laterally and in depth, hence allowing the manufacture of structures with complex geometries, including 3D shapes or structures with varying wall shapes and etch depths [16].

Bearing in mind the complexity of any process involved in the preparation of bio-implants, such as synthesis, micro- and nano-structuring, cell culture, in-vivo, in-vitro assays, in-depth studies are required on the bio-environment–biomaterial interaction mechanism. Furthermore, optimization of the morphology, topography and physico-chemical composition of the contact surface should be carried out.

The aim of the current study is to investigate the optical properties of biopolymer film specimens of chitosan and chitosan with additives of silver nanoparticles. The chitosan/silver nanoparticles (AgNPs) samples are first laboratory synthesized, then modified by femtosecond and nanosecond laser irradiation. The last step is performing analyses in order to investigate their optical properties.

The article is organized as follows – the Introduction section is followed by Section 2 presenting the experimental setup and configuration. In Section 3 results and discussion are provided. Concluding remarks are given in Section 4.

2. Experimental configuration and settings

2.1. Setup configuration

The laser processing is provided by two laser systems – femto- and nano-second lasers. The femtosecond laser system consists of several units – oscillator, amplifier, pumping laser and optical parametric amplifier. The Ti:Sapphire regenerative amplifier has a pulse width of ~35 fs and a pulse energy of ~6 mJ and operates at a 1-kHz repetition rate and 800 nm as the main wavelength. The femtosecond laser is further equipped with a high-precision planar XY stage, a power-meter and optical and opto-mechanical elements. The nanosecond laser system consists of a Nd:YAG laser generating high-power radiation at 1064 nm, 532 nm, 355 nm and 266 nm at a repetition rate of 10 Hz and a pulse duration of 15 ns. Photographs of the femtosecond laser system setup are shown on Figure 1 as an example.

The optical properties are analyzed by a B-150D–BRPL optical microscope (Optica, Italy) and a spectrometer system consisting of a Cary 1E UV-VIS spectrophotometer, an Ocean Optics DH2000 light source and an Ocean Optics HR4000 optical spectrometer. The spectra are obtained with the spectrophotometer operating in a transmission mode. The scanning electron microscopy (SEM) images are recorded by using a JEOL JSM 5510 scanning electron microscope.

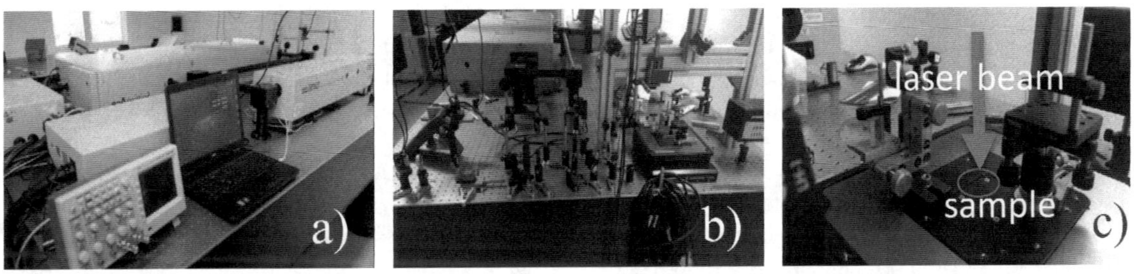

Figure 1. Experimental setup of the femtosecond laser system. Image a) is an overview of the system, image b) shows the optical path elements guiding the laser beam to the sample. Image c) demonstrates the position of the sample on a precise planar stage and the laser beam direction.

2.2. Material

The biopolymer films studied are divided into two groups. The first one is a specimen of chitosan. The chitosan used is of medium molecular weight, Sigma-Aldrich® (Munich, Germany). The chitosan powder is dissolved in a 0.5-mol/L CH3COOH solution and dH2O and heated to 50 °C under continuous stirring for 2.5 h. The second one is chitosan with additives of silver nanoparticles. A solution of silver nanoparticles AgNps–10% Np (v/v) is added to the prepared solution. Suspensions of chitosan-silver blends are prepared with the desired weight ratio of 90% chitosan and 10% silver nanoparticles in 0.5-M acetic acid. The biopolymer films are deposited onto laboratory glass plates and further dried in air. The dimensions of the samples are of 2 cm×2 cm size with 2-mm thickness (Figure 2). Each sample is irradiated and modified by a laser source with controlled sets of laser parameters and then analyzed.

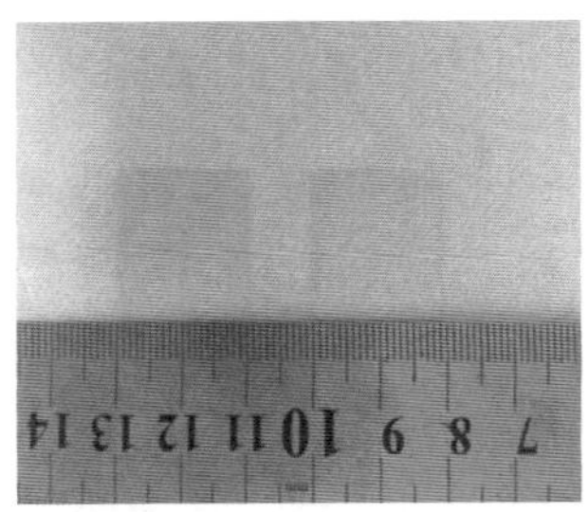

Figure 2. Photo image of chitosan/AgNPs samples.

3. Results and Discussion

The polymer samples are modified under a variety of laser conditions, changing the laser fluence, number of pulses and wavelengths. The laser processing measurements included and discussed in this study are performed by a femtosecond (fs) laser at the wavelength of 800 nm and a nanosecond (ns) laser at 355 nm. The number of laser pulses applied is varied from 1 to 5, which is found to be the optimal condition in this respect [17]. The optimal applied laser fluence in order to avoid ablation or reaching the substrate bottom is found to be 1.2 J/cm^2 for the fs laser and 3.75 J/cm^2 for the ns laser.

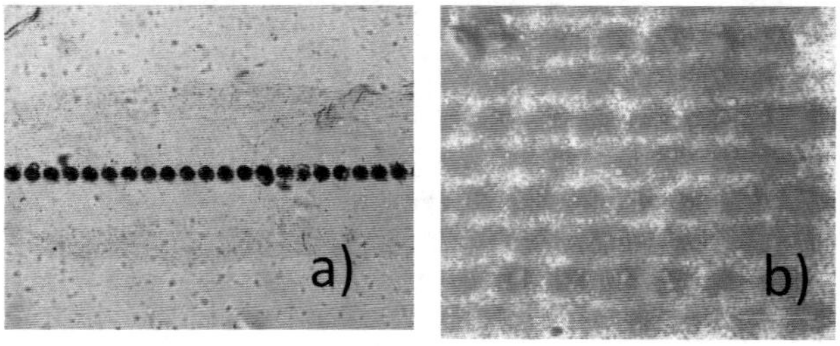

Figure 3. Optical microscope images. Images a) and b) show samples with laser spots grid structures produced by the fs and ns laser, respectively.

Figure 3 shows optical microscope images. The samples on Figure 3 a) and b) exhibit grid structures of laser spots. Each sample is treated by a single laser pulse (per laser spot) of fs and ns laser for a) and b) respectively. The laser fluence for sample a) is $F = 1$ J/cm^2 and for sample b) $F = 3.75$ J/cm^2. The pattern of the laser-irradiated areas and the shape of the laser spots are clearly seen. The laser spot size for the case of the fs laser at wavelength 800 nm is 100 μm and 200 μm for the case of ns laser at wavelength 355 nm.

Figure 4 presents the results of the optical density measurements. The working conditions are as follows – the applied number of pulses is $N = 1$ and 2 at wavelength $\lambda = 355$ nm and laser fluence $F = 3.75$ J/cm^2. The distance between the laser spots is varied. In one case, the spots are adjacent, and in the other, the laser spots overlap. The bottom curve in the graphic in Figure 4 shows the result from an untreated chitosan/AgNPs sample. The top four lines represent measurements of treated samples. The differences in the conditions for these samples concern the number of pulses and the distance (adjacent or overlapping) between the laser spots. It is seen that the optical density changes approximately fivefold after laser irradiation. The other observation is that the samples are very sensitive to changing the distance between the laser spots, i.e., whether they are overlap or are adjacent. The results indicate clearly that by controlling precisely the laser patterning one could easily choose the desired optical density within a certain range.

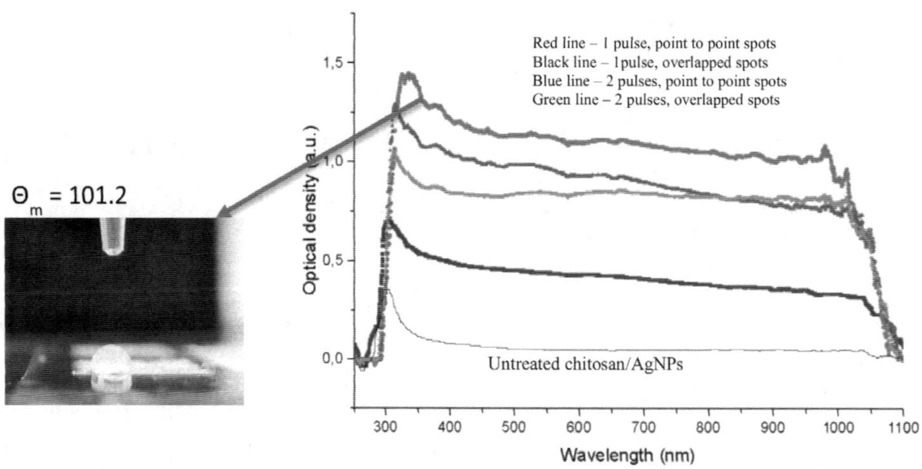

Figure 4. Optical density as a function of the wavelength. Working parameters – wavelength 355 nm, number of ns laser pulses 1 and 2, laser fluence 3.75 J/cm^2. The photograph on the left-hand side of the figure shows a drop sample used to determine the wettability contact angle.

Further, he wettability contact angle (WCA) was also measured. A difference was observed of the WCA values in the treated ($\theta = 101.2°$) and the untreated zones ($\theta = 80°$). Thus, we were able to alter the surface properties from hydrophilic to hydrophobic, which is closely related to potential antimicrobial application of the designed thin film.

IOP Conf. Series: Materials Science and Engineering 1056 (2021) 012002 doi:10.1088/1757-899X/1056/1/012002

The morphology of the samples was monitored by scanning electron microscopy (SEM). Figure 5 presents SEM images of a representative chitosan/AgNPs sample irradiated by a single laser pulse with fluence $F = 3.75$ J/cm^2 at a wavelength of 355 nm. One can clearly see (in the background) a periodic structure of well-defined elliptic shapes with fine circular shapes inside. This pattern is observed in all samples. To the best of our knowledge, such morphological patterning has not so far been reported in the literature. Thus, further detailed and extensive research will be dedicated to explaining this phenomenon.

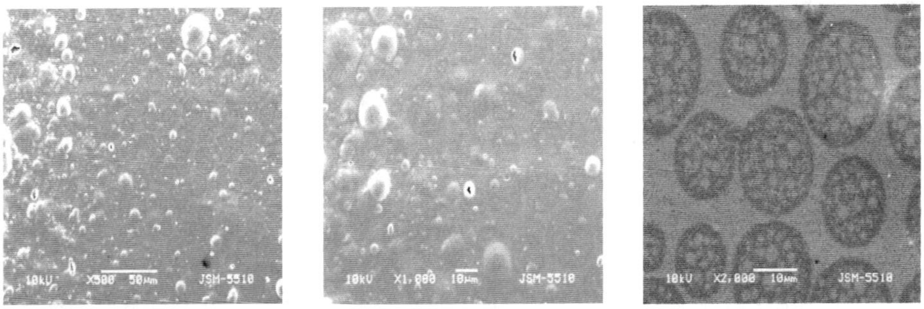

Figure 5. SEM images of chitosan/silver AgNPs sample irradiated with laser fluence $F = 3.75$ J/cm^2 and one laser pulse, at different magnification.

4. Conclusions

The results obtained illustrate the potential of the method of laser irradiation of chitosan/silver AgNPs to obtain various porous modifications depending on the laser parameters applied. The preliminary findings demonstrate that a suitable control of the laser parameters enables one to achieve changes in the optical properties of the biopolymer thin films investigated. Furthermore, adding AgNPs will drastically increase the antimicrobial properties of the thin chitosan films and thus enhance the biocompatibility properties of the 2D matrix created.

As future next steps, we plan to continue investigating the properties of chitosan/AgNPs films with respect to (*i*) their antibacterial activity following established protocols and procedures; (*ii*) the periodic patterning observed by the SEM images.

Acknowledgments

The authors acknowledge the financial support of the Bulgarian National Science Fund under project KP-06-N 38/5 2019.

References

[1] M.M. Rahman, et al. Preparation and properties of biodegradable polymer/nano-hydroxyapatite bioceramic scaffold for spongy bone regeneration, *J. Polym. Eng.* 39(2), 134–142, 2019.

[2] K. Balagangadharan, S. Dhivya, N. Selvamurugan, Chitosan based nanofibers in bone tissue engineering *Int. J. Biol.Macro- mol.,* 104, 1372-1382, 2017.

[3] K. Balagangadharan, S. V. Chandran, B. Arumugam, S. Saravanan, G. D. Venkatasubbu, N. Selvamurugan, Chitosan/nano-hydroxyapatite/nano-zirconium dioxide scaffolds with miR-590-5p for bone regeneration, *Int. J. Biol. Macromol.*, 111, 953, 2018.

[4] Q. Ao, A. Wang, W. Cao, L. Zhang, L. Kong, Q. He, Y. Gong, X. Zhang, Manufacture of multimicrotubule chitosan nerve conduits with novel molds and characterization in vitro, *J. Biomed. Mater. Res. A* 77 11–18, 2006

[5] S.P. Soundarya, A.H. Menon, S.V.Chandran, N. Selvamurugan, Bone tissue engineering: Scaffold preparation using chitosan and other biomaterials with different design and fabrication techniques, *Int. J. Biol. Macromol.*, 119, 1228-1239, 2018.

[6] S. Dhivya, A. Keshav Narayan, R. Logith Kumar, S. Viji Chandran, M. Vairamani, Cell Proliferat., 51, e12408, https://doi.org/10. 1111/cpr.12408, 2018.
[7] L. Zhang, Q. Ao, A. Wang, G. Lu, L. Kong, Y. Gong, N. Zhao, X. Zhang, A sandwich tubular scaffold derived from chitosan for blood vessel tissue engineering, *J. Biomed. Mater. Res. A* 77 277–284, 2006.
[8] S. Husain, K.H. Al-Samadani, S. Najeeb, M.S. Zafar, Z. Khurshid, S. Zohaib,S.B. Qasim, Chitosan biomaterials for current and potential dental applications, *Materials* 10 (6) 602, 2017.
[9] B. Poornima, P. S. Korrapati, Fabrication of chitosan-polycaprolactone composite nanofibrous scaffold for simultaneous delivery of ferulic acid and resveratrol, *Carbohydr. Polym.*, 157, 1741-1749, 2017.
[10] T. A. Hai, R. Sugimoto, Surface modification of chitin and chitosan with poly(3-hexylthiophene) via oxidative polymerization, *Appl. Surf. Sci.*, 434, 188, 2018.
[11] Bazaka, K.; Jacob, M.V.; Crawford, R.J.; Ivanova, E.P. Plasma-assisted surface modification of organic biopolymers to prevent bacterial attachment. *Acta Biomater*. 2011, 7, 2015–2028, 2011.
[12] Cheng, J.-Y.;Wei, C.-W.; Hsu, K.-H.; Young, T.-H. Direct-write laser micromachining and universal surface modification of PMMA for device development, *Sens. Actuators B Chem.*, 99, 186–196, 2004.
[13] M.G. Katsikogianni, Y.F. Missirlis, Bacterial adhesion onto materials with specific surface chemistries under flow conditions. *J. Mater. Sci. Mater. Med.*, 21, 963–968, 2010.
[14] Y. Liu, S. Sun, S. Singha, M.R. Cho, R.J. Gordon, 3D femtosecond laser patterning of collagen for directed cell attachment. *Biomaterials*, 26, 4597–4605, 2005.
[15] R. Murugan, P. Molnar, K.P. Rao, J.J. Hickman, Biomaterial surface patterning of self-assembled monolayers for controlling neuronal cell behavior, *Int. J. Biomed. Eng. Technol.*, 2, 104–134, 2009.
[16] Review Xiaoming Li, Rongrong Cui, Lianwen Sun, Katerina E. Aifantis,YuboFan,Quingling Feng, Fuzhai Cui, and Fumio Watari' *International Journal of Polymer Science*, Volume2014, 2014.
[17] A. Daskalova, S.R. Chandra Nathala, I. Bliznakova, E. Stoyanova, A.Zhelyazkova, T. Ganz, S. Lueftenegger, W. Husinsky, Controlling the porosityof collagen, gelatin and elastin biomaterials by ultrashort laser pulses, *Appl.Surf. Sci.* 292, 367–377, 2014.

Impact of post annealing on the electrical properties of silver nanofilms

Shiva. L. Udachan[1], N. H. Ayachit[1], Udachan. L. A[2], Shivakumar Siddanna[3], Shrishail. S. Kolkundi [4], Ramya S[5]

[1] Departmnt of Physics, Rani Channamma University, Belagavi-591156, Karnataka, India.
[2] S. S. Tegnoor Degree College, Kalaburagi-585105, Karnataka, India
[3] Dept of PG Studies & Research in Physics, Kuvempu University Jnanasahyadri, Shankaraghatta-577451, Shimoga, Karnataka, India
[4] Government First Grade College, Shahapur-585223, Yadgir, Karnataka, India
[5] Shree Sangam Vidya Mandir, Kalaburagi-585104, Karnataka, India

Corresponding author: shivaudachan8@gmail.com

Abstract This paper reports on the growth of nanostructures of silver films prepared by thermal evaporation in vacuum. We focused on the effect of annealing on the silver films' electrical properties. The resistance of the sample was measured by the standard four-probe technique. The electrical resistivity data pertaining to the impact of annealing was analyzed with the help of Fuchs-Sondheimer (F-S) and Mayadas-Shatzkes (M-S) theories.

Keywords: Electrical resistivity, Annealing, Fuchs-Sondheimer (F-S), Mayadas-Shatzkes (M-S) theories.

1. Introduction

Silver has long been utilized for coating mirrors and table ware due to its high light reflecting power. In special cases, silver island films or small particles have been applied as photographic materials, tools and devices to generate localized electric fields or for molecular detection and electric field enhancement [1]. Silver has high electrical and thermal conductivity in bulk form. The electrical properties of the silver films can be improved by altering the film deposition parameters along with the thickness. Therefore, here we address the impact of annealing on the electrical properties of silver films.

2. Experimental Section

Silver nanofilms were fabricated by thermal evaporation in vacuum. The separation between the substrate and the evaporation source was around 0.22 m. The rate of deposition was 0.2 nm/s. Silver of 99.99% purity, Leico Industries, New York, USA, was evaporated onto glass substrates held at room temperature (22 ^{o}C). Prior to deposition, the glass substrates were subjected to a series of cleaning processes, such as using detergent and chromic acid, ultrasonic rinsing in distilled water, baking and finally, ionic bombardment in a vacuum chamber. Silver pieces were placed in a molybdenum boat and heated by a strong electric current. Due to Joule heating, the boat becomes red hot and the silver material evaporation begins. The evaporated silver atoms travel along a straight line in vacuum and condensation of these atoms takes place on the substrate kept above the evaporation source. This gives rise to the formation of silver films. The silver films' resistance was measured by the standard four-probe method. The films' thickness was determined using a digital quartz-crystal thickness monitor housed in the vacuum chamber above the evaporation source.

3. Theoretical Section

Electrical conduction in bulk metals arises from the valence electrons of the constituent atoms. Metallic substances have high thermal and electrical conductivities because of the presence of a large number of free electrons. This was first proposed by Drude [2] in 1900 and later verified by Lorentz [3] in 1904-05. According to Mathiessen's rule [4], the total electrical resistivity [5] of a bulk sample, ρ_B, is equivalent to the sum of resistivities arising from different scattering processes (Maissel, 1970) [5] i.e.,

$$\rho_B = \rho_T + \rho_V + \rho_I \qquad (1)$$

where ρ_T is the component of the resistivity due to the thermal vibrations of the lattice atoms, ρ_V is due to vacancies and ρ_I is due to impurities. An accurate study of the size effect was conducted by F-S by solving Bolztmann's transport equation [6] with appropriate boundary conditions. They derived an expression for the resistivity, ρ, [7]

$$\rho = \rho_0[1 + \frac{3}{8\lambda}(1 - p)],\ \lambda > 0.1 \qquad (2)$$

where ρ_0 is the resistivity of an infinitely thick film, l is the conduction electrons mean free path, t is the thickness of the film. Equation (2) works well over a wide range of $\lambda = \frac{t}{l}$. Further, this expression applies to completely diffuse reflection, whereby the conduction electrons lose their energy in the field direction after collisions with the film surfaces. F-S introduced a specularity parameter, p, representing the part of electrons that are specularly scattered from the film surfaces, so that the electrons contribute to the conductivity even after being scattered off the surfaces. The specularity parameter, p, is a dimensionless quantity having values in the range from 0 to 1, the extremities being a totally diffuse (0) and a totally specular (1) scattering. According to F-S, p is an empirical constant. What causes the diffuse scattering [24] is not yet known; it may be due to the angle at which the carriers are incident on the surface of the film causing scattering and the frequency of collisions. Thus, the total film resistivity can be derived [8-11]. The first term in the Eqn (2) represents the background resistivity, while the second term indicates the resistivity due to the size effect. The F-S theory explains the experimental results of resistivity only for films of larger thicknesses. However, a discrepancy arises between the experimental and the F-S theoretical resistivity data for lower film thicknesses. M-S [12] modified the F-S resistivity equation by incorporating grain boundary scattering, which may be written as

$$\rho = \rho_0\ [1 + \frac{3}{8\lambda}(1 - p) + \frac{3}{2}\alpha' + \dots\dots..], \qquad (3)$$

where α' is the grains boundaries scattering power. M-S assumed a grain size nearly equal to the film thickness up to a certain limit. After this limit, l becomes too small in comparison with the size of the grain boundaries, leading to zero contribution to the total resistivity.

4. Results and Discussion

We annealed the silver films in-situ at a temperature of 100 °C for two hours. There was an appreciable change in the resistance upon annealing along with the other parameters as shown in

Figure 1, which displays the F-S and M-S theoretical results and our experimental curves for both non-annealed and annealed silver films.

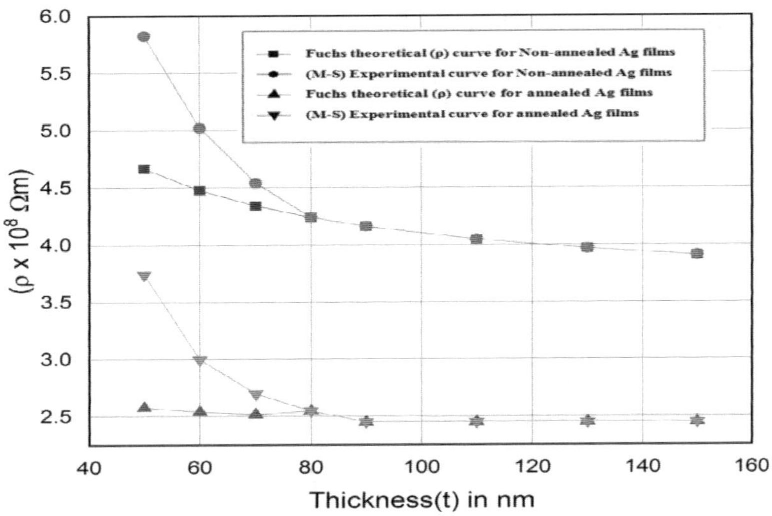

Figure 1. Resistivity (ρ) versus thickness (t) of silver films.

Supplementary Data: Figure 1.

Thickness (t) in nm	Non-annealed (ρ_{Fuchs}) $\times 10^{-8}\ \Omega$ m	Non-annealed (ρ_{exp}) $\times 10^{-8}\ \Omega$ m	Annealed (ρ_{Fuchs}) $\times 10^{-8}\Omega$ m	Annealed (ρ_{exp}) $\times 10^{-8}\ \Omega$ m
50	4.665	5.826	2.58	3.74
60	4.476	5.024	2.542	3.00
70	4.341	4.54	2.517	2.70
80	4.24	4.24	2.55	2.55
90	4.161	4.161	2.45	2.45
110	4.047	4.047	2.45	2.45
130	3.967	3.967	2.45	2.45
150	3.909	3.909	2.45	2.45

Figure 2 is a plot of ($\rho \times t$) *vs* (t) for both non-annealed and annealed silver films. The various parameters estimated from the slopes and intercepts of the two straight lines are listed in Table 1.

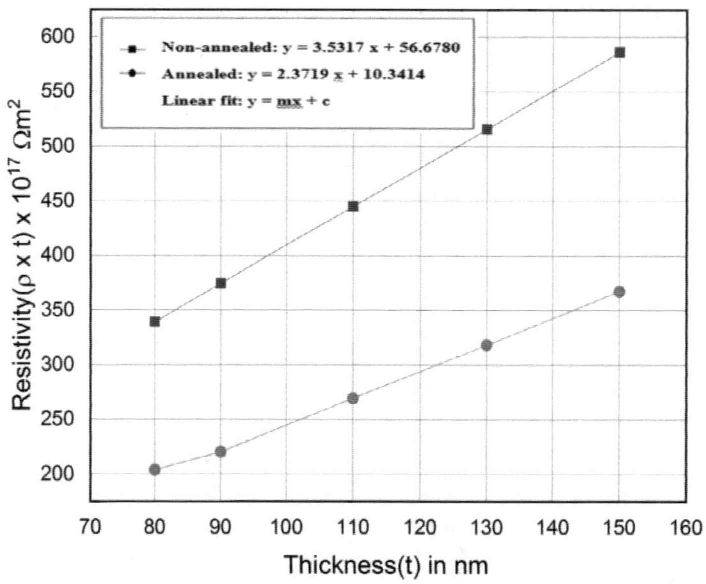

Figure 2. Resistivity ($\rho \times t$) versus thickness (t) for silver films, $t > 80$ nm.

Supplementary Data: Figure 2.

Thickness (t) in nm	Un-annealed $\times 10^{-8}\,\Omega$ m	Annealed $\times 10^{-8}\,\Omega$ m
80	339.22	204
90	374.53	220.5
110	445.17	269.5
130	515.80	318.5
150	586.44	367.5

Table 1. Annealing-based electrical parameters of silver films

Material	Thermal position	ρ_0 ($\times 10^8$) Ω m	$l(1-p)$ nm	l (nm)	p
Silver	Non-annealed Ag film	3.539	42.8	42.8	0
	Annealed Ag film at 100^0C	2.37	11.634	14.54	0.2

Films deposited by evaporation usually harbor a large number of defects, unless great care is taken to avoid them. It is possible to remove these defects by annealing, which generally results in a decrease in the film resistivity [13]. A reduction upon annealing of 20-50 % of the resistance of 24 elements has been reported by Belser [14]. We found that annealing at 100 °C at a pressure of 10^{-6} Torr lowered the annealed film's resistivity about 1.5 times compared with that of the non-annealed film.

5. Conclusions

The resistance of a metal film decreases rapidly on annealing as observed in the cases of silver and other metallic films. The F-S theory fits well our experimental resistivity *versus* thickness data for thicknesses greater than 80 nm. The F-S theory does not consider the grain boundary scattering mechanism, which is predominant in very thin films. The M-S theory explains our experimental results faithfully for annealed and non-annealed films. The annealing improves the electrical characteristics of silver films.

Acknowledgements: One of the authors, Prof. Dr. N. H. Ayachit, acknowledges the Vision Group on Science & Technology, Government of Karnataka, India, for providing the necessary research facility for carrying out the research work and the instruments procured under grant sanctioned bearing GRD No 539.

References

[1] Tominaga Junji 2003 *Journal of Physics: Condensed Matter* **15** R1101.
[2] Drude P 1900 *Annalen Der Physik* **306** 566-613 doi:10.1002/andp.19003060312.
[3] Lorentz H A 1909 *Theory of Electrons: TeubnerVerlag* Leipzig.
[4] Ruei-Cheng Lin, Tai-Kuang Lee, Der-Ho Wu and Ying-Chieh Lee 2015 *Advances in Materials Science and Engineering* **847191** 1-7. http://dx.doi.org/10.1155/2015/847191.
[5] Maissel L I and Glang R 1970 Handbook of Thin Film Technology McGraw-Hill New York 13-4.
[6] Munoz R C and Arenas C 2017 *Applied Physics Reviews* **4** 011102. doi:10.1063/1.4974032.
[7] Larson D C 1971 *Physics of Thin Films* edited by Francombe M.H. and Hoffman, R.W., A.P., Academic Press, New York, **6** 84.
[8] Guowen Ding, César Clavero, Daniel Schweigert and Minh Le 2015 *AIP Advances* **5** 1-12 https://doi.org/10.1063/1.4936637.
[9] Akira Kawazu, Yoshio Saito, Hajime Asahi and Goroh Tominaga 1976 *Thin Solid Films* **37** 261-266. https://doi.org/10.1016/0040-6090(76)90192-9.
[10] JaeSik Jin, JoonSik Lee and Ohmyoung Kwon 2009 *Appl. Phys. Lett.* **95** 052110 https://doi.org/10.1063/1.3202418.
[11] Yusuke Kitaoka, Takeshi Tono, Shinya Yoshimoto, Toru Hirahara, Shuji Hasegawa and Takayuki Ohba 2009 *Appl. Phys. Lett.* **95** 052110. https://doi.org/10.1063/1.3202418.
[12] Mayadas A F and Shatzkes M 1970 *Physical Review* **B1** 1382 DOI:https://doi.org/10.1103/PhysRevB.1.1382.
[13] Leon I Maissel and Reinhard Glang 1970 *Handbook of thin film technology* New York, McGraw Hill 13-31. https://trove.nla.gov.au/version/26289694.
[14] Richard B Belser 1957 *Journal of Applied Physics* **28** 109 https://doi.org/10.1063/1.1722559.

Interfacial compatibilization of PLA and Mg in composites for bioresorbable bone implants

M Ben Abdeljawad, X Carette, R Mincheva, J. Odent, J-M Raquez

Laboratory of Polymeric and Composite Materials (LPCM), Center of Innovation and Research in Materials and Polymers (CIRMAP), University of Mons, Mons, Belgium
E-mail: meriam.benabdeljawad@umons.ac.be

Abstract. In this study, polylactide (PLA)/magnesium (Mg) composites were produced through extrusion and compression-molding. In order to enhance the interfacial adhesion between the hydrophobic matrix and the hydrophilic filler, an amphiphilic PEO-*b*-PLLA block copolymer was used. The morphological study shows an effective improvement of the PLA/Mg interactions following the copolymer addition. Moreover, the surface contact angle test proves the decrease of the PLA hydrophobicity. Thus, a significant influence on the cell adhesion is expected. Furthermore, hydroxyapatite formation in bulk after eight weeks of the immersion in a simulated body fluid (SBF) is also shown, suggesting that the bioactivity will be noticeably improved. However, a decrease in mechanical properties and cell adhesion is observed.

1. Introduction

Stainless steel, cobalt-based alloys and especially titanium-based alloys have been the key materials for bone implants since the 1980s. However, these permanent implants present some disadvantages, such as side effect from long-term use: implant erosion and partial resorption, cortical bone erosion and fragmentation due to the high Young's modulus of the implant requiring second/multiple surgical interventions [1]. Thus, alternatives to permanent metal implants already available on the market, such as bioresorbable implants, have been developed and continue to gain popularity. Together with the requirements for any implant (biocompatibility and suitable mechanical properties), the bioresorbable implants include biodegradability and bioactivity so as to achieve adequate strength and stiffness maintained during a required time interval, allow cell adhesion and proliferation, degrade and resorb for bone healing until its complete regeneration. For these reasons, the polylactide (PLA) is among the biodegradable and biocompatible polymers of choice [2]. However, PLA possesses low mechanical properties (Young's modulus (GPa) = 3-4) compared to bone performance Young's modulus (GPa) = 7-30), uncontrollable and relatively slow degradation time (> 6 months) and acidic degradation products (causing local inflammation issues) [3]. To successfully remediate this issues, (nano)fillers have often been incorporated to polymers. Among all, Mg present good osteoconductivity, high mechanical properties (compared to monolithic polymers) and lower degradation times (< 1 month) [4, 5]. Bearing in mind the published results, we propose here a combination of PLA and Mg. However, the poor interfacial adhesion between the hydrophobic PLA and the hydrophilic Mg is expected to negatively affect the matrix/filler (composite) properties. One of the possibilities to enhance the compatibility between a polymer and a filler involves modification of the filler surface using compatibilizing agents [6, 7]. We report on the testing of an amphiphilic copolymer to form a new interface between the matrix and the filler, thus improving the interfacial adhesion of PLA/Mg biocomposites and the composite's biological activity.

2. Preparation of composites

The amphiphilic diblock copolymer poly(ethylene oxide-*block*-polylactide), (PEO-*b*-PLLA) was synthesized via bulk ring opening polymerization (ROP) of L-lactide initiated by a PEO with a fixed chain length of 5000 g/mol. [8] The amount of copolymer was fixed to 10 wt.% in PLA. The filler amount was 5 wt.%, 10 wt.% and 15 wt.% of Mg and composites were prepared via extrusion and shaped into cylinders (12-mm length and 4-mm diameter) and rectangular specimens of 60×12×3 mm^3 (length×width×thickness) via compression-molding.

3. Results and discussions

3.1 Morphology, surface hydrophilicity and dynamic mechanical analysis

3.1.1 Morphology

The morphology of all the composites was visualized using scanning electron microscopy (SEM) to evaluate the influence of PEO-*b*-PLLA on the PLA/Mg interfacial adhesion. The samples were cryofractured and covered with gold prior to observation. As seen in **Figure 1** (A), the poor interactions between the hydrophobic PLA matrix and hydrophilic Mg filler resulted in cavities left by the Mg particles after cryofracture. This behavior is considered as typical for incompatible composites with low interfacial adhesion between the matrix and the filler [5]. In contrast, the presence of the amphiphilic copolymer allowed an interestingly good compatibility between PLA and Mg as evidenced by the absence of cavities and the smooth and homogenous surface after criofracture (**Figure 1** (B).

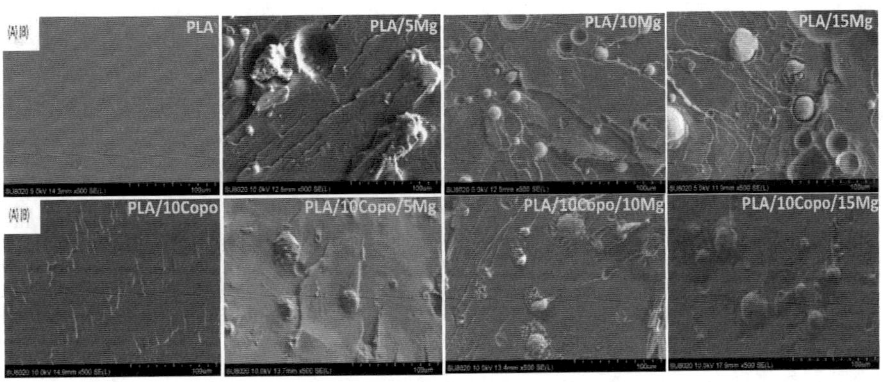

Figure 1. SEM images in bulk of (A) PLA/xMg and (B) PLA/10Copo/xMg composites (x = 0, 5, 10 and 15 wt. %).

3.1.2 Surface hydrophilicity

Together with the surface-morphology, surface hydrophilicity is known to tailor the bioactivity of implants. [8] Therefore, the influence of the copolymer on the PLA/Mg surface hydrophilicity was evaluated using water contact angle (WCA) analysis. Interestingly, the addition of PEO-*b*-PLLA showed a positive effect on the composite wettability. Indeed, the amphiphilic copolymer is able to enhance the hydrophilicity of the PLA/Mg composites by selective surface localization of the hydrophilic PEO blocks. [9] Thus, a decrease in the WCA from the first set (without copolymer) to the second one (with copolymer) was observed (table 1).

Table 1. Water contact angle values of PLA/xMg and PLA/10Copo/xMg composite surfaces versus Mg amount (x = 0, 5, 10 and 15 wt.%).

	Water Contact Angle	
Mg amount (wt.%)	**PLA/xMg**	**PLA/10Copo/xMg**
0	93	81
5	84	78
10	81	76
15	79	72

3.1.3 Dynamic mechanical analysis

Dynamic mechanical analyses (DMA) are often used to assess the mechanical properties of samples as a function of the temperature. The results obtained for all composites and the neat PLA are presented in table 2

. As seen, an improvement in the storage moduli (E') was reached at 10 wt.% of Mg and 37 °C for the first set (PLA/xMg). Concerning the PLA/10Copo/xMg set, the E' reproduces the same behavior of the first one with a decrease in the E' values. In reality, the E' decrease indicated the plasticizing effect of the PEO present in the PEO-b-PLLA copolymer [10]. After the copolymer addition, a significant drop in the loss factor tanδ was detected when the Mg amount increased. According to the literature, the decrease of tanδ is related to the highest intensity of interface adhesion [11]. Consequently, the composites reinforced with an amount of 10 wt.% of Mg filler will be more closely investigated in the following section.

Table 2. DMA results (storage modulus E' and loss factor tanδ) of PLA/xMg and PLA/10Copo/xMg composites (x = 0, 5, 10 and 15 wt. %) at 37 °C.

	PLA/xMg		PLA/10Copo/xMg	
Mg amount (wt.%)	**E'(GPa)**	**tanδ**	**E'(GPa)**	**tanδ**
x=0	2.0	59	1.5	57
x=5	2.1	62	1.7	53
x=10	2.5	61	2.1	54
x=15	2.3	58	1.6	50

3.2 In vitro degradation study

In order to evaluate the bioactivity of the composites, an in vitro degradation test was performed by soaking the cylinders in the SBF at 37 °C and at pH 7.4. Prior to this test, the samples were taken out of the SBF and rinsed several times with distilled water, wiped off and dried in nitrogen. The visual aspect of the PLA/10Mg and PLA/10Copo/10Mg composites revealed the formation of a white friable layer on the composites surface. Its thickness for the PLA/10Copo/10Mg was found to also increase over time. This might be explained by the formation of hydroxyapatite 'HAp' (the major natural bone component) based on Mg degradation. [12] According to the literature [13], Mg degrades in SBF mainly through reactions with water by producing Mg cations (Mg^{2+}) and hydroxide anions (OH^-) and releasing hydrogen gas. The Mg^{2+} and OH^- form the magnesium oxide layer. Then, Ca^{2+} and Mg^{2+} react with phosphate to form small quantities of calcium phosphate ($Ca_3(PO_4)_2$ and magnesium phosphate ($Mg_3(PO_4)_2$

Figure). Indeed, some cracks and pitting corrosion were observed in the composites (figure 2). This behavior is more pronounced in PLA/10Copo/10Mg compared to PLA/10Mg composite because of the hydrophilicity of PEO. In order to confirm the chemistry of the white layer, EDX analysis was performed on the composites surface and in bulk (cross-section). The data obtained revealed that after a period of eight weeks HAp was only formed at the surface of PLA/10Mg (Ca/P=1.6), while the mineral was present in the entire bulk of the PLA/10Copo/10Mg. Interestingly, the content of Ca and P was more prominent in the presence of copolymer in bulk. Indeed, the Ca/P ratio was 1.64 reached after eight weeks of degradation versus 1 for the PLA/10Mg composite. Thus, the copolymer-containing samples seem more beneficial for new bone tissue formation during the bone-healing process [14].

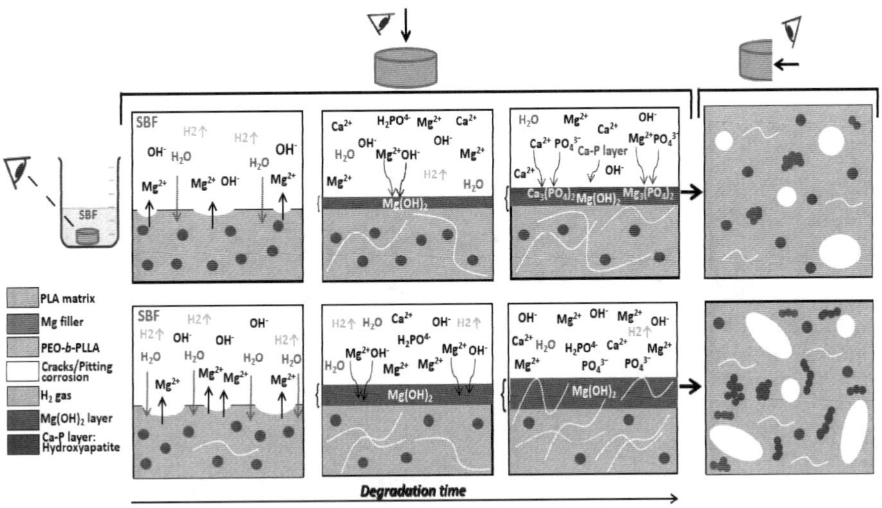

Figure 2. Degradation mechanism of the PLA/10Mg and PLA/10Copo/10Mg in the SBF.

4. Conclusion

As was shown above, the presence of an amphiphilic copolymer enhances the PLA/Mg interfacial adhesion and decreases the PLA hydrophobicity via creating a completely new interface between PLA and Mg. In fact, it is assumed that the more the surface is hydrophilic, the more the hydrophilic bone-like layer adheres. Moreover, HAp – the major component of natural bone – was formed in bulk only after eight weeks of immersion in the SBF for the PLA/10Copo/10Mg. However, the copolymer presence decreased the mechanical properties of the composite-based materials. In this regard, research is in progress to develop composites that meet the mechanical properties of natural bone by using novel co (polymers) to improve the PLA/Mg adhesion.

Acknowledgments

The authors thank the M-ERA.NET for supporting POLYMAGIC Project and the European Regional Development Fund and the Walloon Region for supporting the LCFM-BIOMAT project and the Bioprofiling platform. Jean-Marie Raquez is a FRS-FNRS senior scientific researcher.

References

[1] Sandra Carolina C, Rosario B, Marcela L and José Luis G C 2017 *Handbook of Composites from Renewable Materials* 1-8 (Spain, Wiley online library) 109-143
[2] Jean-Marie R, Youssef H, Marius M and Philippe D 2013 *Progress in Polymer Science* 38 1504-1542
[3] Hong C, Yue Z, Xuan L, Jiao M, Feng X, Chenglin C, Li T and Jing B 2018 *Composites*

Science and Technology 162 198-205

[4] Changli Z, Hongliu W, Jiahua N, Shaoxiang Z and Xiaonong Z 2017 *Composites Science and Technology* 147 8-15

[5] Sandra Carolina C, Rosario G, Marcela L, Rosario B and José Luis G C 2016 *Acta Biomaterialia* 32 348-357

[6] Nikushi Y, Dilpreet B and Sreekala B 2010 *Composites Part A: Applied Science and Manufacturing* 107 315-325

[7] Ana F M, Marcela L, Rosario B, José Luis G C and Begoña F 2020 *Additive Manufacturing* 33 101142

[8] Jean Marie R, Deborah B, Zofia L, Olivier P, Alexandra B, Jeroen E, Jan S and Philippe D 2011 *Biomacromolecules* 12 692-700

[9] Tiziano S, Monica O H, Elisabeth E, Josep A P and Melba N 2014 *Materials Science & Engineering C* 38 55-62

[10] Livia L, Pablo C T, Gustavo A and Maria F 2017 *RSC Advances* 7 161-172

[11] Thatiana B, Maria Odila H C and Herman V ECCM 2012 *Composites at Venice, Proceedings of the 15th European Conference on Composite Materials* no 24-28

[12] Wahaaj A, Ali M, Min-Gu H and Seung-Hwan C 2019 *Composite Structures* 210 914-921

[13] Rong-Chang Z, Xiao-Ting L, Li-Jun L, Shuo-Qi L and Fen Z 2016 *Journal of Materials Science and Technology* 32 437-444

[14] Hong C, Jiao M, Xuan L, Feng X, Chenglin C, Chao G and Jing B 2019 *Acta Biomaterialia* 98 125-141

IOP Conf. Series: Materials Science and Engineering 1056 (2021) 012005 doi:10.1088/1757-899X/1056/1/012005

Atmospheric plasma: a simple way of improving the interface between natural polysaccharides and polyesters

X Carette[1], R Mincheva[1], M Herbin[2], X Noirfalise[2], T C Nguyen[3], P Leclere[3], T Godfroid[2], H Kerdjoudj[4], O Jolois[5], M Boudhifa[6], J Raquez[1]

[1] Laboratory of Polymeric and Composite Materials (LPCM), University of Mons, Place du Parc, 23, B-7000, Mons, Belgium
[2] Laboratory of Chemistry of Plasma-Surface Interaction (ChIPS), University of Mons, 23 Place du Parc, B-7000 Mons, Belgium; Materia-Nova Research Center, Parc Initialis, B-7000 Mons, Belgium
[3] Laboratory for Chemistry of Novel Materials (CMN), University of Mons, Place du Parc 23, Mons 7000, Belgium
[4] EA 4691 Biomatériaux et Inflammation en Site Osseux (BIOS), Rue du Maréchal Juin 1, Reims, France
[5] Centre du Textile Belge (CENTEXBEL), Rue du Travail 5, Grâce-Hollogne, Belgique.
[6] CRITT-MDTS, Boulevard Jean-Delautre 1, Charleville-Mézières, France

jean-marie.raquez@umons.ac.be

Abstract. In this study, a novel simple and fast method was developed of immobilizing a natural polysaccharide (chitosan) on a polylactide (PLA) surface. In order to overcome the lack of surface adhesion between the chitosan and the PLA due to incompatibility issues, atmospheric pressure plasma was used. The plasma activation was able to modify the PLA surface by generating surface chemistry proper for stabilizing the chitosan layer. The activation of the surface was monitored through water contact angle (WCA) measurements and X-ray photoelectron spectroscopy (XPS). Indeed, the introduction of new functionalities lowers the hydrophobicity of the PLA surface allowing an effective spreading and immobilization of the chitosan layer. The presence of chitosan on the polyester's surface was proven by Fourier transform infrared spectroscopy (FTIR).

1. Introduction

Tissue engineering (TE) is a multidisciplinary field of science focused on the regeneration of damaged living tissues [1]. Knowledge in chemistry, physics, engineering, and clinical sciences is combined to solve the problems related to tissue loss and organ failure via the design of a 3D scaffold with specific properties, allowing the cells to reproduce and form a natural implant. As the materials used for the 3D scaffold design have to combine multiple properties, finding a single material component that can fulfill all the requirements is highly improbable [2]. Thus, multiple combinations between polymers (inert, biodegradable, bioactive) and particles (metallic, ceramic, polymeric) are often studied. In terms of polymers, for example, polyesters are widely used in 3D scaffold preparation due to their biocompatibility and biodegradability. More particularly, the polylactic acid (PLA) has a long history of uses and has gained the US Food and Drug Administration (FDA) approval for clinical use. Despite the polymer biocompatibility, biodegradability, and bioresorbability, the adhesion of cells to the surface of the hydrophobic PLA remains a major challenge.[3], which limits the polyester uses in the biomedical field [4].

Many ways to improve the surface-cell compatibility can be found in the literature [4–7]. One of them relies on the introduction of bio(macro)molecules onto the surface of the polyester. In this respect, chitosan gains interest for its outstanding biological properties related with cell adhesion and antimicrobial and wound dressing properties. However, the incompatibility between the PLA and the chitosan chemistries implies the use of a prior chemical modification of one or both polymers to allow efficient coating or grafting. Usually, chitosan modification or covalent crosslinking is used to anchor the biopolymer onto the PLA surface. However, this significantly alters chitosan's biological activity and limits the success of the final scaffolds. Modifying the PLA seems more appropriate, but when performed in the polymer bulk this might alter the PLA functions. Therefore, other methods limited to surface modification are under extensive development. Amongst all, atmospheric plasma has already demonstrated its capability of modifying the surface chemistry of polymers and preserving their bulk properties [8].

In this work, a microwave atmospheric plasma torch was used to improve the adhesion of chitosan to the PLA surface. The plasma creates new functionalities for interacting with chitosan and an interface for its adhesion. The activation of a PLA surface was investigated by WCA and XPS. The presence of chitosan was also investigated by FTIR and WCA.

2. Experimental

2.1 Materials

The polyester used in this work is an amorphous PLA from INGEO (PLA 4060D). It possesses a T_g around 55 °C and no melting temperature. The PLA pellets were dried under reduced pressure overnight before thermo-compression at 180 °C to obtain a square sample of 4 cm² and a thickness of 500 µm. The chitosan is commercially available (Sigma) from shrimp with a deacetylation degree of 75% and a molar mass of 300 kg/mol. Acetic acid (ACS reagent, 99.7% purity) was supplied by Sigma and used as received. The chitosan was used as received and solubilized in an aqueous medium with the help of acetic acid (1 wt%). The final concentration of chitosan was 1 mg/ml.

2.2 Samples preparation

The plasma treatment was realized using an IONJET (a microwave plasma torch from IONICS, Belgium) with a generator set at a frequency of 2.5 GHz and a power of 30 W. Ar (99.9% purity, Air Liquide) was used as a carrier gas at a flow of 2 L/min. The torch was placed vertically at 1 cm off the sample surface fixed on an XY table moving at a speed of 8 mm/sec. A single cycle was performed and the samples were then dip-coated for 10 s in the chitosan solution. The as-prepared samples were dried in ambient air overnight.

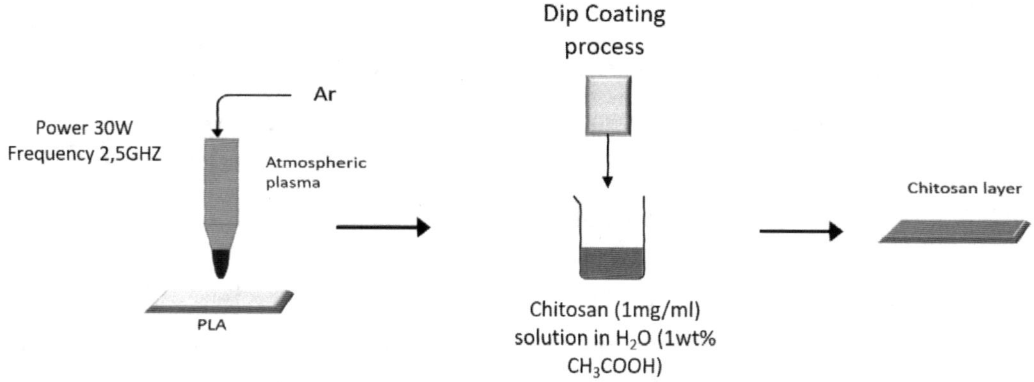

Figure 1 Schematic representation of the PLA surface modification and chitosan deposition.

IOP Conf. Series: Materials Science and Engineering 1056 (2021) 012005 doi:10.1088/1757-899X/1056/1/012005

2.3 Analyses

The FTIR spectra were recorded on a Bruker Tensor 27 spectrometer in attenuated total reflectance mode with a 4-cm^{-1} resolution. The background and sample spectra were obtained in the wavenumber range 600 – 4000 cm^{-1}. The water contact angle (WCA) measurements were implemented using a DSA Mk2 drop shape analysis system in a static mode at 25 °C by a sessile drop method. For each sample, at least five measurements on different surface locations were averaged. The XPS analyses of all samples were performed using an ESCA-5000 (Physical Electronics) Versa Probe system. The following X-ray settings were used: beam-size diameter 200 μm; beam power 50 W; voltage 15 kV. The pressure in the analysis chamber was typically 2×10^{-6} Pa. The XPS data were collected using monochromatic Al Kα radiation at 1486.6 eV. The photoelectrons were collected at a take-off angle of 45° (normal detection) to the surface normal. Survey (general) spectra were taken with a 187.85 eV pass energy, a 0.5 eV step and 40 ms/step, and the high-resolution spectra, with a 23.5 eV pass energy, a 0.2 eV step and 50 ms/step. The atomic compositions were derived from the high-resolution peak areas using the photoionization cross-sections calculated by Scofield, corrected for the dependence of the escape depth on the kinetic energy of the electrons and for the analyzer transmission function of our spectrometer. The atomic compositions were derived from the peak areas after a Shirley background subtraction.

3. Results and discussion

In order to fully assess the different steps of the process (Figure 1), *(i)* the plasma conditions were defined to prevent any modification of the PLA bulk or over-degradation of the polyester surface; *(ii)* the surface chemistry of the PLA was completely studied; and *(iii)* the chitosan layer deposition was investigated.

The change of the water contact angle (WCA) is a clear indication of a change in a polymer surface chemistry [9]. When applied to the untreated and the plasma-activated PLA surfaces, this easily accessible and well-known technique showed a sharp decrease of about 30° in WCA (Figure 2). Indeed, while the WCA of PLA was close to the typical value of 85° [10], after plasma activation it dropped to 55°. This result attests to a decrease in the surface hydrophobicity [7] and is related to imparting oxidized functions onto a polyester surface [11]. These newly formed reactive species form hydrophilic groups (hydroxyl or carboxylic) that will be finally used to interact with chitosan [12].

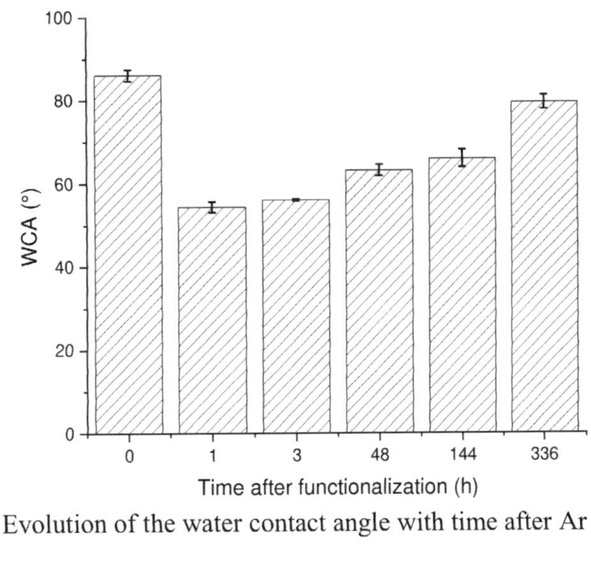

Figure 2 Evolution of the water contact angle with time after Ar treatment.

However, this change in the surface hydrophobicity was not permanent, as an increase of the WCA was observed with time within two weeks. This phenomenon is due to a relaxation process of the surface known as hydrophobic recovery [13]. The period for effective chitosan deposition was then defined to be of several hours and the coating process was realized directly after the plasma activation.

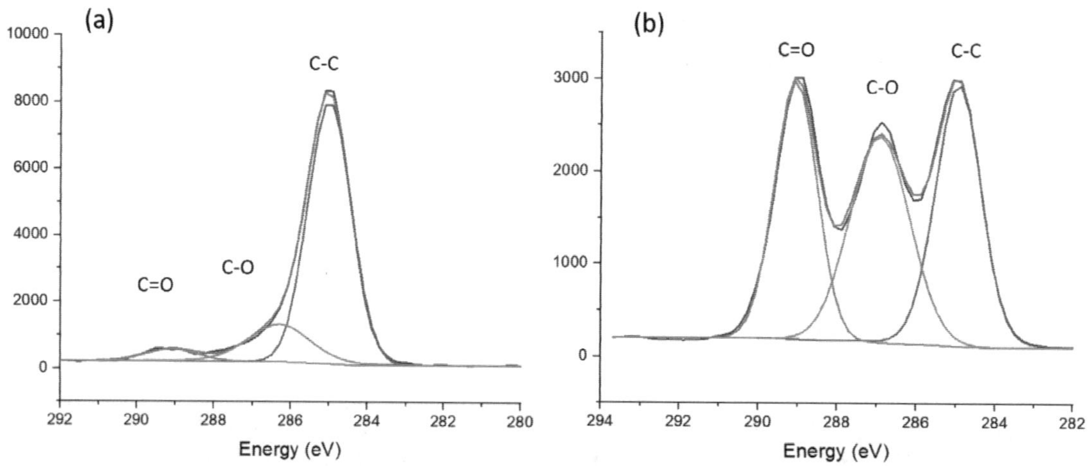

Figure 3 C1s analysis of (a) neat PLA and (b) activated PLA

To confirm and determine the plasma-created chemistry, the surface composition of the samples was studied by XPS. Figure 3 (a)-(b) shows the deconvoluted C1s core-level spectra of the pristine and the activated PLA, respectively. The C1s spectrum of the pristine PLA surface contained three peaks: 285 eV (-C-H/C-C) (80%), 286.3 eV (C-O-C/C-OH) (16%), and 289.2 eV (O-C=O) (4%) [14]. However, the proportion of C-O to C=O increased when the sample was subjected to plasma activation. The surface was highly enriched in oxygen, the oxygen content rising from 10 to 40 atomic %. These results are associated with the formation of oxidized functions on the surface as concluded from the WCA results [15]. The presence of these new hydrophilic functions will have two purposes. They will allow obtaining a better spreading of the chitosan solution during the dip-coating process[16] and stabilization of the chitosan coating on the PLA surface.

At the next step, chitosan layer deposition was performed by short-time dip-coating to prevent any possible PLA surface alterations. The chitosan presence at the surface was then studied by FTIR (Figure 4). The characteristic bands of polyester can be seen on the reference (black curve), the CH_2 and CH_3 vibrations around 2800 $cm^{-1,}$ and the carbonyl around 1700 cm^{-1}. As a control, one sample was dip-coated without any plasma activation (red curve). Its spectrum is almost overlapping the reference curve highlighting the necessity of the plasma activation process in the stabilization of this chitosan layer. For the plasma-activated and dip-coated sample (act+dpc), two major observations can be made. First, the peaks of the polyester are still present on the curve, which means that the layer of chitosan is thinner than the depth of analysis (i.e., 2 μm) or discontinuous. As several spots of the sample showed this behavior, the first hypothesis seems more plausible. Second, three new peaks appear on the spectrum, two relatively sharp bands of amide I and II respectively at 1650 cm^{-1} (stretching of the C=O group hydrogen bonded to N–H of the neighboring intra-sheet chain) and 1580 cm^{-1} (NH_2 bending), and a wide peak corresponding to the amine and hydroxyl functions around 3300 cm^{-1} [17] attributed to the presence of chitosan. These groups can be assigned to the OH of the sugar unit and the amine of the lateral group of the D-glucosamine unit of the chitosan.

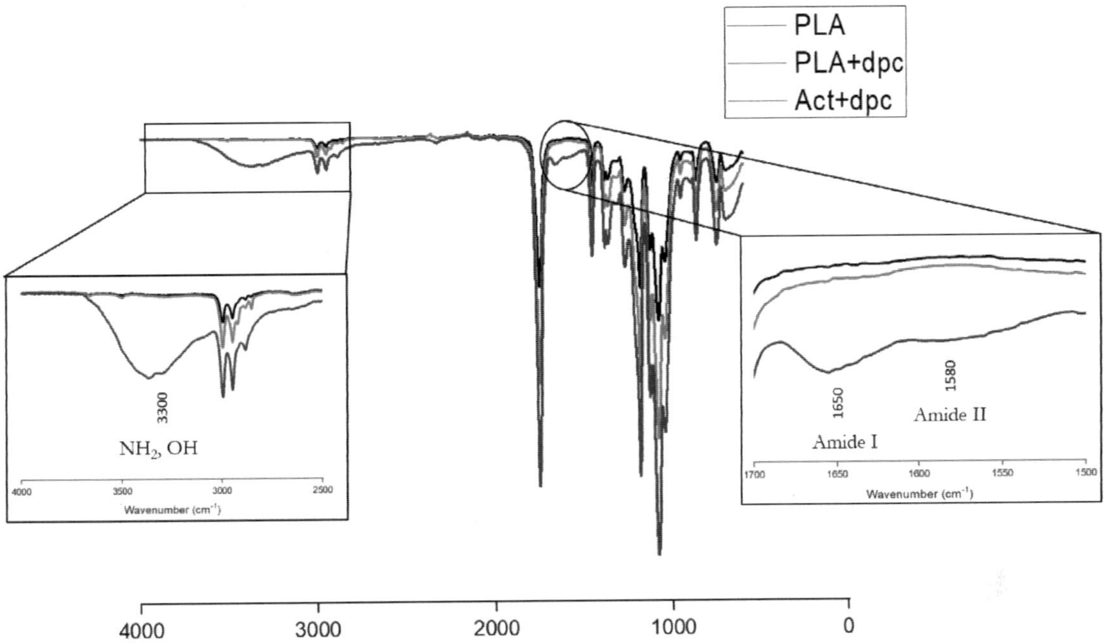

Figure 4 FTIR spectra of neat PLA (black), a dip-coated PLA (dpc, red) and a plasma-activated + dip-coated PLA (act+ dpc, blue)

4. Conclusions

In this work, a new technique of improving the immobilization of a polysaccharide, chitosan, on a polyester surface was developed. Thanks to the plasma technology employed, the polyester surface could be modified to possess enough functions to allow the immobilization of the chitosan as proved by FTIR, contact angle measurements and XPS. This technique exhibits several advantages, as it is fast, efficient, and could be easily adapted to treat 3D objects in reduced time compared to the usual chemical methods [18]. However, several issues are still to be dealt with. The morphology changes of the surface during the process and a chemical investigation of the final layer to assess the interactions involved in the stabilization will be part of a future publication.

5. Acknowledges

Authors thank INTERREG IV program, BIOMAT-FEDER, FEDER LCFM, and RW for supporting TEXTOS project. CIRMAP is grateful to the European Commission and the Wallonia Region (FEDER Program), and OPTI2MAT program of excellence. Jean-Marie Raquez is a FRS-FNRS senior scientific researcher.

References

[1] Jagur-Grodzinski J 2006 *Polym. Adv. Technol.* **17** 395–418.
[2] Grøndahl L, Jack K S and Goonasekera C S 2017 *"Composite materials for bone repair"* Second Edi, Elsevier Ltd.
[3] Armentano I, Dottori M, Fortunati E, Mattioli S and Kenny J M 2010 *Polym. Degrad. Stab.* **95** 2126–2146.
[4] Nakagawa M, Teraoka F, Fujimoto S, Hamada Y, Kibayashi H and Takahashi J 2006 *J. Biomed. Mater. Res. A.* **77** 112–118.

[5] Yang J, Bei J and Wang S 2002 *Biomaterials* **23** 2607–2614.
[6] Saranya N, Saravanan S, Moorthi A, Ramyakrishna B and Selvamurugan N 2011 *Biomed. Nanotechnol.* 7 238-244.
[7] Tamada Y and Ikada Y 1993 *Polymer (Guildf).* **32** 2208–2212.
[8] Leroux F, Campagne C, Perwuelz A and Gengembre L 2009 Surf. Coat. Technol. **203** 3178–3183.
[9] Neuendorf R E, Saiz E, Tomsia A P and Ritchie R O 2008 *Acta Biomater.* **4** 1288–1296.
[10] Darie-Niţă R N, Vasile C, Stoleru E, Pamfil D, Zaharescu T, Tarţău L, Tudorachi N, Brebu M A, Pricope G M and Dumitriu R P 2018 *Materials (Basel).* **11** 1825.
[11] Ikada Y, Suzuki M and Tamada Y 1984 *"Polymer Surfaces Possessing Minimal Interaction with Blood Components"* Polymers as Biomaterials, in: S.W. Shalaby, A.S. Hoffman, B.D. Ratner, T.A. Horbett (Eds.), Springer US, Boston, MA, 135–147.
[12] Wang R, Shen Y, Zhang C, Yan P and Shao T 2016 *Appl. Surf. Sci.* **367** 401–406.
[13] Mortazavi M and Nosonovsky M 2012 *Appl. Surf. Sci.* **258** 6876–6883.
[14] Nazmul M, Rigout M, Yeates S G and Carr C 2014 *Dye. Pigment.* **103** 168–174.
[15] Kostov K G, Nishime T M C, Castro A H R, Toth A and Hein L RO 2014 *Appl. Surf. Sci.* **314** 367–375.
[16] Dowling D P, Miller I S, Ardhaoui M and Gallagher W M 2010 *J. Biomater. Appl.* **26** 327–347.
[17] Zhu A, Zhang M, Wu J and Shen J 2002 *Biomaterials* **23** 4657–4665.
[18] Ma Z, Mao Z and Gao C 2007 *Colloids Surfaces B: Biointerfaces* **60** 137–157.

Electron-beam surface alloying of Ti substrates with Ta films

S Valkov[1], M Ormanova[1], R Bezdushnyi[2], and P Petrov[1]

[1] E. Djakov Institute of Electronics, Bulgarian Academy of Sciences, 72 Tsarigradsko Chaussee, 1784 Sofia, Bulgaria
[2] Faculty of Physics, St Kliment Ohridski University of Sofia, 5 James Bourchier blvd., 1164 Sofia, Bulgaria

E-mail: stsvalkov@gmail.com

Abstract. We report electron-beam surface alloying of Ti substrates with Ta. The Ti substrates were covered by a Ta coating with a thickness of 2 μm. The samples were alloyed by a scanning electron beam, with the beam deflection geometry in the shape of the infinity symbol (∞). Two experiments were implemented, where the width of the infinity symbol was kept constant at 8 mm, while the height was 8 mm for the first experiment and 4 mm for the second one. The phase composition of the specimens produced was studied by X-ray diffraction (XRD). The microstructure and chemical composition were investigated by scanning electron microscopy (SEM) and energy-dispersive X-ray spectroscopy (EDX), respectively. The results showed that the phase composition consisted of a double phase structure of α' martensitic and of beta phases. Some amount of pure Ta remained in the alloyed layer produced by the first experiment. Also, the distribution of the alloying element into the substrate was significantly more homogeneous in the case of the second one.

1. Introduction

Nowadays, titanium and its alloys are widely used for the manufacturing of implants, artificial joints and in other branches of modern medicine due to titanium's biocompatibility, excellent corrosion resistance, and static and fatigue strength [1-3]. However, some limitations due to the poor wear resistance should be mentioned. Since the surface of the discussed materials is exposed during the implants use, the discussed limitations can be overcome by an appropriate surface-treatment technology.

A number of methods for surface modification exist, such as thin-film deposition [4], surface treatment and alloying by high-energy fluxes (i.e. electron [5, 6] and laser [7] beams), etc. Presently, alloying by a flux of accelerated electrons is widely studied due to the shorter process time and the possibility of controlling precisely the technological conditions. In this technology, the electrons interact with the surface of the alloyed material where the technological conditions are so optimized as to reach its melting temperature and form a melt pool. The alloying elements are introduced into the molten material; after its solidification, a surface alloy is formed with significantly improved functional properties in comparison with the initial material [8].

Currently, the formation of surface alloys and coatings of the Ti-Ta system is considered as very promising for various biomedical applications due to the improved corrosion resistance and biocompatibility [9, 10]. The Ti-Ta materials are known as a non-toxic beta alloy, which exhibits high specific strengths, toughness, and fracture resistance. Furthermore, this material exhibits a high-temperature shape memory effect, while its hardness and Young's modulus are much closer to that of

IOP Conf. Series: Materials Science and Engineering 1056 (2021) 012006 doi:10.1088/1757-899X/1056/1/012006

human bones. The authors of [11] have demonstrated the formation of Ti-Ta coatings on TiNi substrates by pulsed electron-beam melting of the film/substrate system. The results showed that the coatings exhibited better elastoplastic properties in comparison with the substrate. It was found that in the depth the hardness and depth recovery ratio decrease, while the plasticity increases. Golkovski et al. [12] have studied the possibilities of the formation of Ti-Ta coatings on VT1 titanium substrates by atmospheric electron-beam surface alloying. The results showed the formation of surface layers with thicknesses of about 2 – 2.5 μm and an α(α′)+β structure. An increase in the Ta content led to a significant increase in the tensile strength and the corrosion resistance.

However, the capabilities of the electron-beam surface alloying of pure Ti substrates with Ta films, as well as the influence of the process conditions on the structure of the Ti-Ta coatings, are currently less well investigated. Therefore, this study aims to investigate the possibilities of forming Ti-Ta coatings by alloying pure titanium substrates with a Ta film via electron-beam surface alloying with a continuously scanned electron beam. Further, the influence is also discussed of the alloying process conditions defined by the geometry of the electron-beam deflection on the structure of the produced layer.

2. Experimental part

In our study, Ti-Ta surface alloys were produced on commercially pure Ti substrates by using DC magnetron sputtering to deposit Ta films with a thickness of about 2 μm. The deposition process took place in Ar atmosphere with a working pressure of 8×10^{-2} Pa. The discharge voltage was 440 V, the discharge current was 1 A, and the deposition time was 1 hour. The samples were then electron-beam surface alloyed in an Evobeam Cube electron-beam unit. A scheme of the electron beam alloying is presented in fig. 1 (a). During the alloying process, the accelerating voltage was 60 kV, the electron beam current was 25 mA, the speed of the specimen motion was 5 mm/s, and the electron-beam scanning frequency was 1 kHz. It should be noted that the electron-beam equipment used for the experiments allows one to implement a number of beam deflection geometries, such as line, circle, ellipse, parable, infinity, etc. For the needs of these experiments, a continuously scanned electron beam was used, where the scanning figure had the form of the infinity symbol (fig. 1 b). In this case, a significant overlap of the beam trajectory takes place, leading to an increase in the melt pool lifetime. The dimension of the scanning figure was varied to realize two experiments, namely, the width of the infinity symbol was kept constant at 8 mm, while the height was chosen to be 8 mm for the first experiment and 4 mm for the second one (Figure 2).

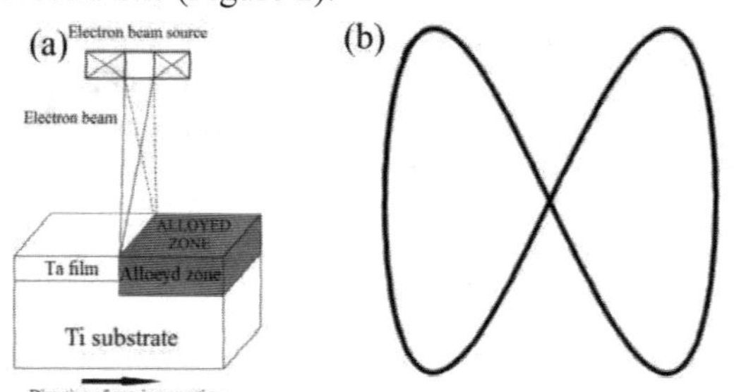

Figure 1. Electron-beam surface alloying of a Ti substrate with a Ta film a) a scheme of the process; b) electron-beam scanning figure.

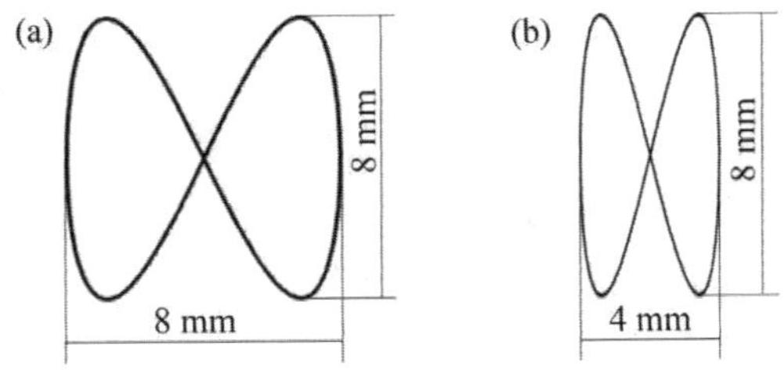

Figure 2. Dimensions of the electron-beam scanning figure a) for the first experiment; b) for the second experiment.

The phase composition of the samples was characterized by X-ray diffraction (XRD) with CuKα (1.54 Å) radiation. The experiments were carried out in a symmetrical Bragg-Brentano (B-B) mode at a 2θ scale within the range from 30° to 75°, a step of 0.1° and a counting time of 10 sec per step.

The layers produced were studied by means of a scanning electron microscope (SEM) equipped with an EDX detector, which integrates a true standardless analysis with P/B ZAF corrections (Z being the atomic number correction factor; A, the X-ray absorption correction factor; F, the fluorescence correction factor). During the experiments, back-scattered electrons were used.

3. Results and discussion

Fig. 3 presents the experimentally obtained XRD results. The phase composition of the obtained surface alloys consists of a double phase structure of hexagonal close-packed (hcp) α' martensitic and of body-centered cubic (bcc) beta phases. Additionally, the diffraction pattern for the first experiment exhibits peaks of pure Ta, meaning that some amount of the tantalum film has not been dissolved into the Ti matrix. The diffraction peaks of the beta phase and Ta are almost overlapped because both compounds have the same crystal structure (body-centered cubic bcc) and very similar lattice parameters [13]. The identification of the discussed phase was done according to ICDD (International Center for Diffraction Data) database, for hcp α' (ICDD PDF # 44–1294); for bcc β (ICDD PDF # 44–1288); for Ta (ICDD PDF # 04-0788). The formation of non-equilibrium phases such as the metastable martensitic phase in the system of Ti-Ta is due to the very high cooling rate during electron-beam surface alloying [14]. During the alloying process, the temperature reaches the values of the transformation from α to β. The subsequent transformation from beta to alpha is difficult due to the very high cooling rate [15]. At the same time, the Ti-Ta materials can be retained in a metastable state at room temperature due to the beta stabilizing effect of the tantalum. The authors of [16] have studied the structure and properties of Ti-Ta alloys formed by selective laser melting, and their results showed that the phase composition of Ti-Ta consists of a double-phase structure of α'martensitic and beta phases. Also, the same authors [16] showed that the amount of β phase rises with an increase in the Ta content. These statements are in agreement with our results obtained by XRD pointing to the formation of the double-phase structure of hcp α' martensitic and bcc beta phases.

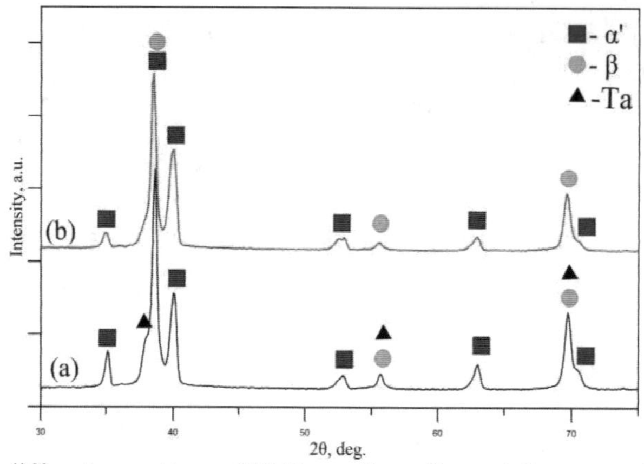

Figure 3. X-ray diffraction pattern of Ti-Ta surface alloys a) first experiment; b) second experiment

Figure 4 a) and b) presents a cross-sectional SEM image of the Ti-Ta surface alloy formed by the first and second experiment, respectively. The chemical composition of the specimen was studied by EDX at several points (for the surface alloy obtained in the first experiment), and areas (for the surface alloy obtained in the second experiment), which are marked in figure 4; the results are summarized in Table 1. Figure 4 (a) demonstrates that the structure of the surface alloy produced in the first experiment is strongly inhomogeneous, with the thickness varying from about 8 μm to 15 μm. The results show that a significant amount of undissolved Ta is present on the top of the formed layer. However, in the deeper parts, the structure contains Ti-Ta. The results obtained by SEM/EDX are in agreement with those from the XRD investigations pointing to the formation of a Ti-Ta surface alloy with a significant amount of undissolved Ta. The structure of the specimen obtained via the second experiment is shown in figure 4 b). The alloyed zone is marked as zone A, while the substrate is indicated as B. It is obvious that the formed layer is much more uniform in comparison with that of the first experiment. The alloying element is distributed significantly more homogeneously into the Ti matrix, and no, or a very small amount of, undissolved Ta fraction is visible. The results for the chemical composition, which are summarized in Table 1, show that a surface alloy in the system of Ti-Ta was successfully formed. It is found that the alloyed layer consists of Ti–18 wt% Ta. Such Ti-Ta alloys are considered promising materials for modern biomedicine [16].

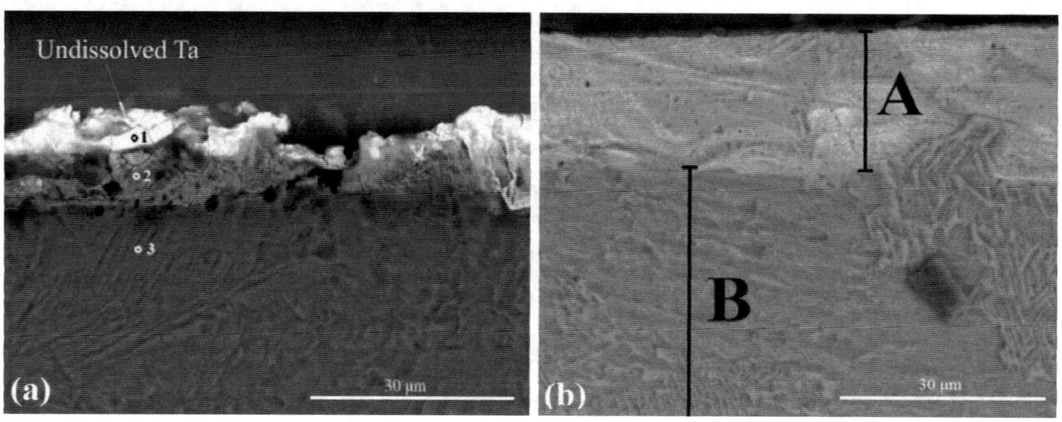

Figure 4. Cross-sectional SEM image of the Ti-Ta surface alloy formed in the a) first experiment and b) second experiment

Table 1. Chemical composition of Ti-Ta surface alloy formed in the first and second experiment

	Ti, wt%	Ta, wt%
Point 1	9.7±0.4	90.3±3.0
Point 2	70.5±1.9	29.5±1.0
Point 3	100.0±2.6	0
Alloyed zone (A)	81.5±2.3	18.5±0.8
Substrate (B)	100±3.5	0

The results obtained in this study show the possibility of forming Ti-Ta-based surface alloys by means of electron-beam surface alloying of a Ti substrate with a Ta film. The influence was studied of the dimensions of the electron-beam scanning figure (defined by the trajectory of the electron-beam scanning) on the structure of the alloyed layers. During the experiments, the scanning figure was in the form of the infinity symbol (fig.1 b), where a significant overlap of the beam trajectory exists leading to an increase in the lifetime of the melt pool. The dimension of the scanning figure was varied, with the width of the infinity symbol kept constant at 8 mm, while the length was chosen to be 8 mm for the first experiment and 4 mm for the second one (figure 2). It was found that in the second experiment the alloying element is homogeneously distributed within the Ti matrix, without undissolved Ta fractions, in contrast with the results obtained in the first experiment. In the case of a larger scanning figure (first experiment), the heat input is significantly lower in comparison with the other one. This leads to a lower melt pool temperature, and thus, a low miscibility between the alloying element and the substrate [17]. Moreover, the use of a smaller electron-beam scanning figure (second experiment) leads to a much higher heat input and sufficient mass transport for melt homogenization. These statements are completely in agreement with the results of the present study where the smaller electron beam scanning figure during the alloying process leads to the formation of a much more homogeneous structure of the surface alloy.

4. Conclusions

The results obtained in this study show the possibility of producing Ti-Ta based surface alloys by means of electron-beam surface alloying of a Ti substrate with a Ta film. The influence was studied of the dimensions of electron-beam scanning figure (defined by the trajectory of the electron beam) on the structure of the alloyed layers. During the experiments, the scanning figure was an infinity symbol. The dimension of the scanning figure was varied – the width of the symbol was kept constant at 8 mm, while the length was chosen to be 8 mm for the first experiment and 4 mm for the second one. It was found that the phase composition of the surface alloys produced consists of a double phase structure of α' martensitic and beta phases, while some amount of pure Ta exists in the alloyed layer obtained in the first experiment. The distribution of the alloying element into the substrate was significantly more homogeneous in the case of the second experiment. The conclusion is drawn that the heat input is of major importance for melt homogenization of the surface alloys formed by selective electron-beam alloying, which can be controlled by the dimension of the electron-beam scanning figure.

Acknowledgments: The financial support is acknowledged of the Bulgarian National Science Fund, project No KP-06-M37/5.

References

[1] Elia C, Lima J, Valiev R and Meyers M 2008 *JOM (J. Occup. Med.)* **60 (3)** 46–49
[2] Jia M, Zhang D, Liang J and Gabbitas B 2017 *Metall. Mater. Trans.* **48 (4)** 2015–2029
[3] Chaves J, Olmos L, Jimenez O, Bouvard D, Rodriguez E and Florez M 2017 *Powder Metall.* **60 (4)** 257–266.

[4] Petrov P, Dechev D, Ivanov N, Hikov T, Valkov S, Nikolova M, Yankov E, Parshorov S, Bezdushnyi R and Andreeva A 2018 *Vacuum* **154** 264-271
[5] Petrov P 2010 *J Phys Conf Series* **223** 012029
[6] Zhang C, Lv P, Xia H, Yang Z, Konovalov S, Chen X and Guan Q 2019 *Vacuum* **167** 263–270
[7] Almeida A, Carvalho F, Carvalho P and Vilar R 2012 *Mater Sci Eng C* **32 (5)** 1190-1195
[8] Petrov P 1997 *Vacuum* **48** 49-50
[9] Martel F, Mirza R, Rodriguez G, Baltes L and Tierean M 2013 *Metal. Int.* **18** 18–22
[10] Capellato P, Riedel N, Williams J, Machado J, Popat K and Claro A 2013 *Engineering* **5(9)** 707-713
[11] Meisner S, Yakovlev E, Semin V, Meisner L, Rotshtein V, Neiman A and D'yachenko F 2018 *Appl Surf Sci* **437** 217-226
[12] Golkovski M, Bataev I, Bataev A, Ruktuev A, Zhuravina T, Kuksanov N, Salimov R and Bataev V 2013 *Mater Sci Eng A* **578** 310-317
[13] Grevey D, Vignal V, Bendaoud I, Erazmus-Vignal P, Tomashchuk I, Daloz D and Sallamand P 2015 *Mater Des* **87** 974-985
[14] Sing S, Yeong W and Wiria F 2016 *J Alloy Compd* **660** 461-470
[15] Fischer M, Joguet D, Robin G, Peltier L and Laheurte P 2016 *Mater Sci Eng C* **62** 852-859
[16] Zhao D, Han C, Li Y, Li J, Zhou K, Wei Q, Liu J and Ghi Y 2019 *J Alloy Compd* **804** 288-298
[17] Zhao H, Li J, Zheng Z, Wang A, Zeng D and Miao Y 2016 *Surf Coat Technol* **286** 303-312

Roughness and mechanical properties of electron-beam surface modified and TiN/TiO_2-coated Ti6Al4V alloy

M P Nikolova[1], M Ormanova[2], D Tonev[1] and P Petrov[2]

[1] Faculty of Mechanical and Manufacturing Engineering, A. Kanchev University of Ruse, 8 Studentska Str., Ruse, Bulgaria
[2] E. Djakov Institute of Electronics, Bulgarian Academy of Sciences, 72 Tsarigradsko Chaussee, Sofia, Bulgaria

Abstract. In this study, we used PVD deposition of TiN/TiO_2 coatings on polished and electron-beam-treated Ti6Al4V alloy to explore the changes in the surface roughness and mechanical properties of the coated systems. After the electron-beam treatment (EBT), the average surface hardness and roughness (R_a) increased from 323±5.62 $HV_{0.2}$ and 0.14±0.004 μm up to 387.5±9.33 $HV_{0.2}$ and 1.58±0.05 μm, respectively. After coating with 3.7-μm thick TiN/TiO_2, the average S_a roughness and nanohardness of the film on the polished substrate reached 0.87±0.1 μm and 13.05±2.07 GPa while that on the EBT was equal to 1.57±0.2 μm and 9.02±2.15 GPa, respectively. However, the comparison of the coefficient of friction (COF) evolution of the substrates and the coated specimens indicated a COF decrease by about 0.18±0.03 for the coated EBT alloy as compared with the polished and EBT substrates, whose average COF were 0.45±0.04 and 0.38±0.06, respectively. A high average COF (0.68±0.16) was registered for the coated as-received alloy because of the gradual and complete wearing off of the film during the test. These results indicate that the combination of initial EBT of the Ti6Al4V alloy with PVD deposition of a TiN/TiO_2 coating could substantially improve the roughness and tribological properties of the coated systems.

1. Introduction

Experiments with animals have shown that the body accepts better the rough surface of implants because of the improved tissue adherence to roughened grafts [1]. A rough surface can be produced by machining, chemical, electrochemical, electrical discharge processes, or by irradiation with high-energy fluxes (e.g., electron or laser beams). Among these, the electron-beam treatment is characterized by beain a non-oxidizing process with precisely controlled energy input and process parameters that benefit the improvement of the surface structure, roughness, and other properties. The process of irradiation triggers fast heating and cooling that induce structural transformations, melting of the surface and changes in the surface roughness.

The most widely recommended titanium alloy for orthodontic and orthopedic implants today is Ti6Al4V. However, it suffers from a lack of quick and effective osseointegration and degradion due to corrosion in the aggressive body environment that releases harmful V and Al ions [2]. This issue can be addressed by applying superficial treatment, such as wet or spray processes, chemical treatments, or deposition of coatings for osseointegration. The PVD-deposited TiN/TiO_2 film appears to be such a bioactive coating that has been proved to enhance the corrosion resistance and improve cells viability, adhesion, and bone mineralization [3, 4]. However, biocompatible coatings can only be considered safe if possessing suitable mechanical properties that can make them functional. Inappropriate hardness and elastic modulus and a high coefficient of friction could lead to a low bio-fixation of the material and failure of the implant in service. Lower Young's modulus values of hard coatings close to that of the cortical bone (17 GPa [5]) can be considered beneficial since a large elastic modulus mismatch between implant and bone may cause stress shielding and bone resorption under load [6]. It is widely recognized that the substrate material plays an important role in determining the coatings' wear resistance and mechanical properties [7]. This paper reports our studies on the effect of electron-beam treatment of Ti6Al4V substrates on the mechanical response of TiN/TiO_2 coatings by using AFM, nanoindentation and ball-on-wear tests.

2. Experimental procedure

Samples with chemical composition 6.22% Al, 3.57% V, 90.38% Ti (in wt %) and dimensions of Ø22×4 mm were cut by electro-erosion from pieces of Ti6Al4V alloy. The as-received (AR) samples were grounded and finely polished. Electron-beam treatment (EBT) was carried out by a Leybold Heraus (EWS 300/15–60) electron-beam installation. The following technological parameters were applied: linear scanning, electron beam current I = 30 mA, acceleration voltage U = 52 kV, speed of sample motion v = 2 cm/s, electron-beam scanning frequency f = 850 Hz. The samples surfaces were ultrasonically cleaned with absolute ethanol, acetone, isopropanol, rinsed with distilled water for 5 min, and dried. The deposition of TiN (by vacuum cathodic arc technology at 340 °C) and overlaying TiO_2 (by glow plasma discharge at 300 °C) has been described in detail elsewhere [8].

The surface morphology of the untreated and of the coated samples was assessed by scanning electron microscopy (SEM, LYRA I XMU, Tescan). Micro-Vickers hardness measurements were conducted on the top of each substrate (polished and EBT) with a load of 200 g and dwell time of 15 s using a ПМТ-3 microhardness (ПОМО) tester. The distance between the indents was 20 μm. The nanohardnness of the coatings was investigated using a nanomechanical tester (Bruker, Billerica MA) which applied a load of 25 mN at 48 indentations with a spacing of 80 μm.

The surface topography was analyzed using a contact profilometer (Mitutoyo SJ 201 P). The samples were evaluated quantitatively for R_a (average roughness) after a tenfold scanning with a cut-off value of 0.8 mm in a direction perpendicular to the grooves caused by the local melting. The surface topology was characterized by atomic force microscopy (Q-Scope™ 250/400 Nomad™ commercial AFM system, Ambios Technology Inc). The measurements were taken by a 10-nm radius silicon tip on each coating over 80×80 μm scan areas. The coefficient of friction (COF) was determined by sliding wear study performed on a ball-on-flat tester (CSEM tribometer, normal load 2 N, linear speed 100 mm/s, time of testing 300 s) with a sliding Ø6-mm ball coated with Cr. The sliding rate and time were chosen so as to initiate wear of the coatings. The tests were conducted in air at room temperature.

3. Results and discussions

The representative SEM image of the as-received (figure 1a) Ti6Al4V alloy shows a lamellar initial structure. By using backscattered electron imaging, the α-phase features were identified by their darker color since they contain aluminum atoms of lower density, whereas the lighter color of the β-phase features is associated with the predominant presence of vanadium atoms of higher density. The wavy surface morphology obtained after the EBT is seen in figure 1b. After the EBT, the average surface hardness and roughness (R_a) increased from 323±5.62 $HV_{0.2}$ and 0.14±0.004 μm up to 387.5±9.33 $HV_{0.2}$ and 1.58±0.05 μm, respectively. When observed at a higher magnification (figure 1c), it is evident that the equiaxial grains on the re-melted surface contain the martensite phase only and no parental microstructure is present. The diffusionless transformation associated with needle formation is a result of the fast cooling above β-transus.

A cross-section micrograph of cathodic-arc-evaporated TiN/TiO_2 coatings deposited on the surface of both as-received and EBT alloys is presented in figure 1d. The high-current density at the arc spot forms not only metal ions, electrons, and neutral atoms, but also agglomerates, or macroparticles, which are seen on the surface and the cross-section parts of the film. Because of the high bias-voltage (Us = –250 V) that increases the deposition flow, the concentration of drops is decreased since purification and activation occurs due to ions of evaporated material spraying the surface by [9]. After the thin transition layer of pure Ti between the coating and the substrate, the nitride layer displays a dense columnar structure with no inter-columnar voids. The thinner top TiO_2 layer displays finer pyramidal grains seen on the surface of the droplet phases. The whole coating thickness measured in figure 1d is 3.72 μm.

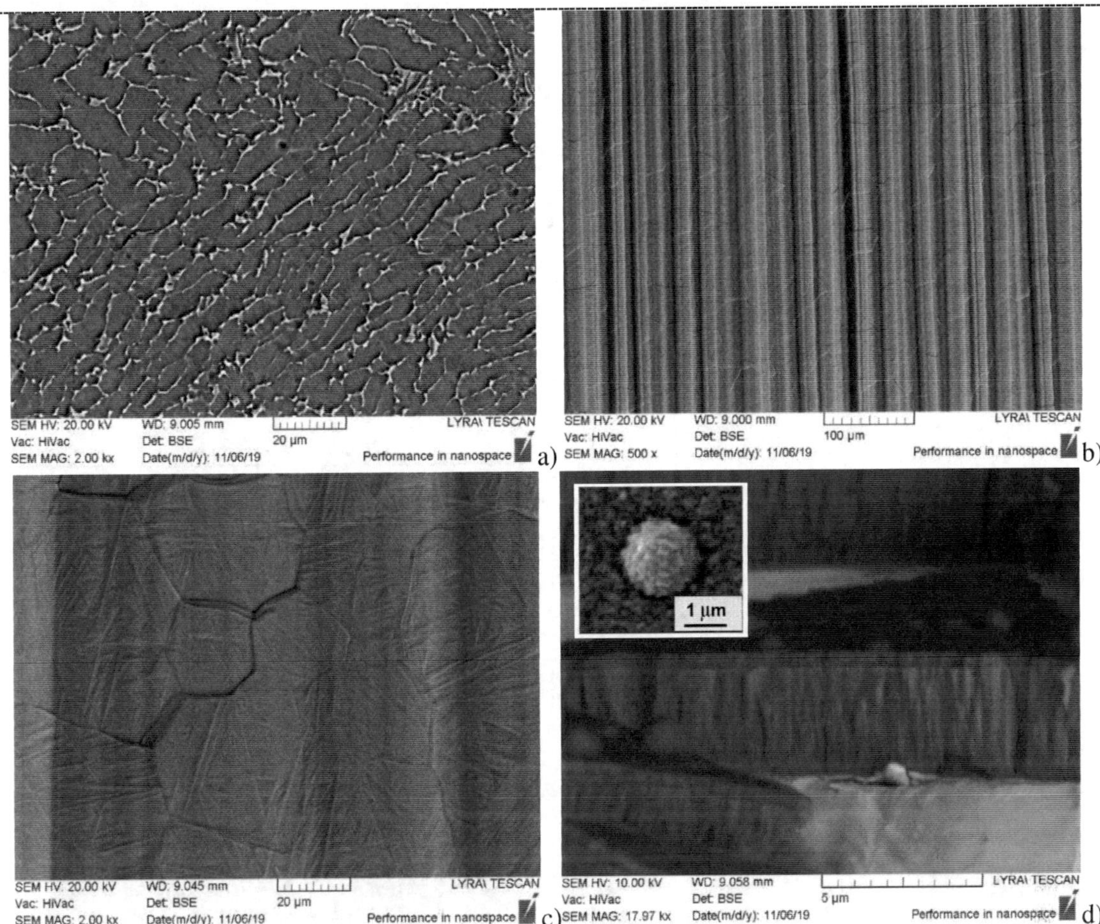

Figure 1. SEM micrographs of representative areas of the a) polished AR alloy; b) EBT substrate material; c) remelted structure of the EBT alloy at higher magnification; d) cross-section of PVD deposited TiN/TiO_2 coating.

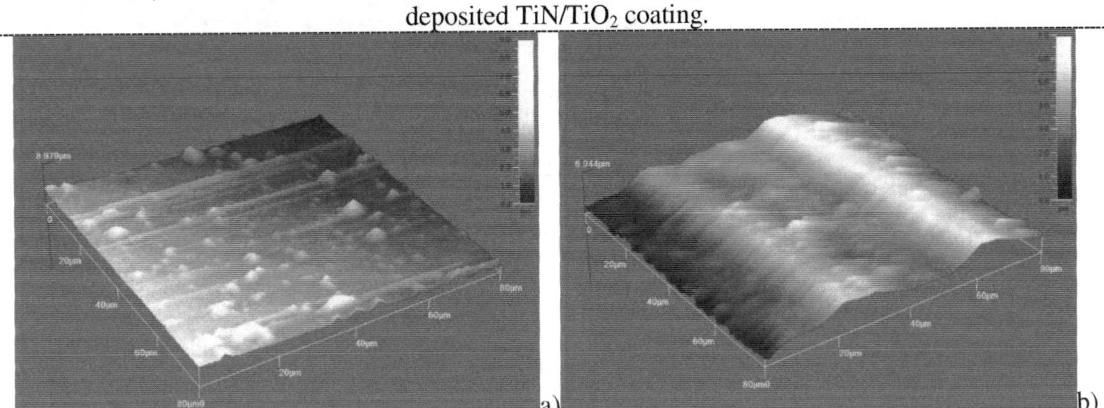

Figure 2. AFM images of the films deposited on polished (a) and EBT (b) alloy.

3D AFM images of the coatings deposited on the surface of the polished and the EBT substrates are shown in figures 2a and 2b, respectively. The average S_a roughness of the film on the polished substrate reaches 0.87±0.1 μm, while that on the EBT sample is 1.57±0.2 μm. As seen in figure 2, the measured difference is attributed to the initial roughness of the substrates, because the PVD coatings replicate the surface topography of the substrates. The small increase in the S_a values of the coated polished sample as compared with the non-coated substrate is due to the presence of macroparticles within the film (figure 2a).

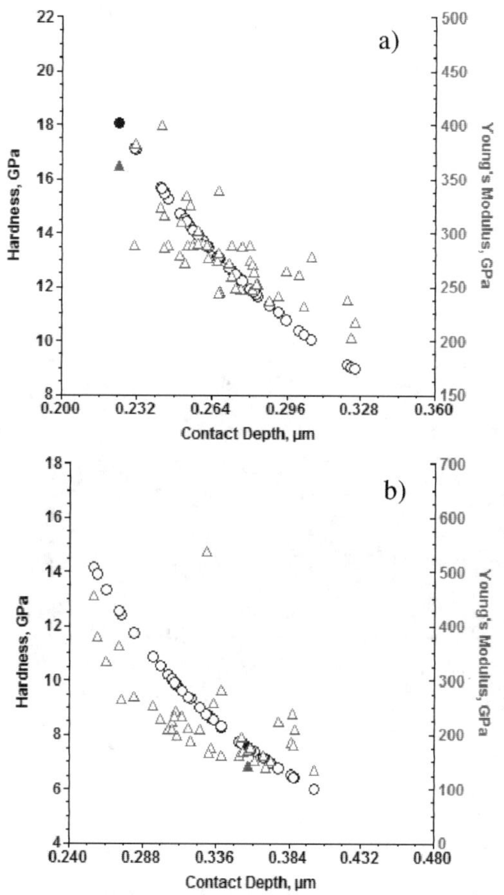

Figure 3. Hardness and Young's modulus as a function of the indentation contact depth of TiN/TiO_2 coatings deposited on polished (a) and EBT (b) samples.

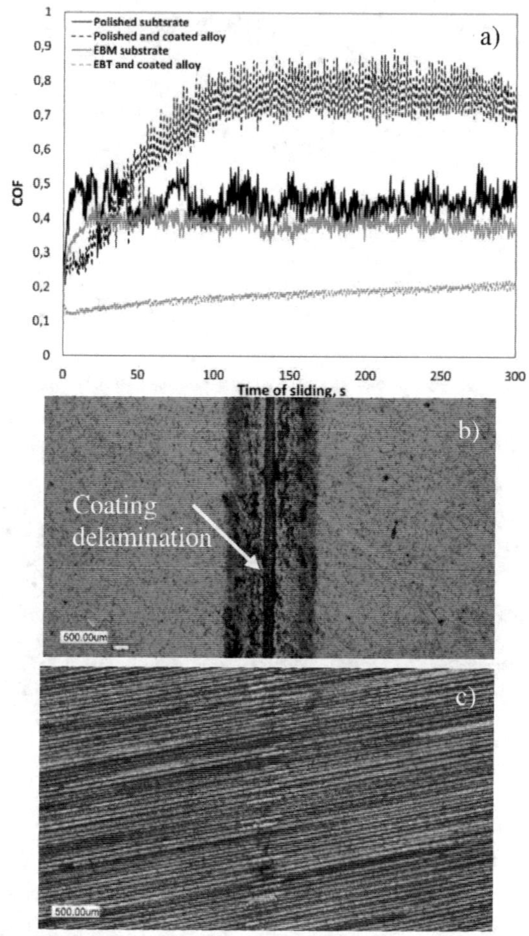

Figure 4. Friction coefficients (a) during sliding of substrates and of TiN/TiO_2 coatings on polished (AR) and EBT (AR850) alloy versus sliding distance; OM images of wear track areas of the coated polished (b) and EBT (c) alloy.

The results from measuring the hardness and the elastic modulus are shown in figure 3. The average nanohardness values for the coatings deposited on the polished and EBT samples were found to be equal to 13.05±2.07GPa and 9.02±2.15 GPa, respectively, whereas those of the Young's modulus were equal to 279.5±39.2GPa and 225.9±85.2 GPa, respectively. The drop in the hardness value of the coatings deposited on the EBT alloy could be attributed to the higher surface roughness values [10], because the flat samples exhibit better surface properties. It is seen that the standard deviation of the hardness and modulus values increase with the roughness. Bearing in mind the coating thickness, at a load of 25 mN the penetration depth is less than 10% of the film thickness and the substrate effect is negligible. The hardness decreases gradually with the increase in the indentation displacement under an identical loading force and varied indentation positions. The large contact depths should be a result of defects in the coatings, such as voids, droplets, or other microstructure inhomogeneities under the indenter.

Figure 4 plots the evolution of the coefficients of friction (COF) of the bare and of the coated samples with the sliding time. After the initial running period, the friction coefficients of the substrates remained almost constant during the sliding tests. Because of the higher surface hardness and roughness of the EBT Ti6Al4V alloy, its average COF was lower (0.38±0.06) as opposed to that of AR and polished alloy (0.45±0.04). Regarding the coated polished sample, one can see an intermediate region of increased COF which rises to the maximum steady-state value suggesting that events occurred corresponding to the formation of deep wear debris. This fact can be attributed to the soft substrate material that is unable to withstand the friction force and, to a lesser extent, to coating delamination (figure 4b). In spite of the higher roughness of the EBT alloy, the higher substrate hardness and the characteristics of the mechanical contact resulted in a substantially lower average COF (0.18±0.03) for the EBT and the coated alloy. The micrographs of the wear tracks (figure 4c) show the delamination of some areas at the coated peaks because of the adhesive wear mechanism.

4. Conclusions

TiN/TiO_2 coatings with a thickness of about 3.7 μm were fabricated using arc PVD on polished and electron-beam surface modified Ti6Al4V alloy. Nanoindentation tests showed different hardness of the coatings deposited on polished and electron-beam modified alloys at similar indentation depths, whereas the response was not dependent on the underlying substrate. Both the substrate roughness and the coating topography influenced the hardness value response measured. The COF of the coated samples was dominated by the plastic deformation of the substrate material. Since the as-received material has a lower hardness, the friction was dominated by ploughing, while that of the coated electron-beam modified Ti6Al4V indicated a more durable and protective behavior. The lower Young's modulus values of the coating on the electron-beam modified alloy that are close to that of the cortical bone (17 GPa [11]) and its better tribological properties make the proposed surface treatment appropriate for further application in the biomedical field.

Acknowledgments

The authors would like to thank Ilian Tzvetkov for depositing the coatings and Emil Yankov for cutting the samples and participating in the development of the glow plasma-discharge system.

References

[1] Major S, Cyrus P and Hubálovská M 2017 *IOP IOP Conf. Series: Mater Sci & Eng.* **175** 012053 doi:10.1088/1757-899X/175/1/012053

[2] Khadija G, Saleem A, Akhtar Z, Naqvi Z, Gull M, Masood M, Mukhtar S, Batool M, Saleem N, Rasheed T, Nizam N, Ibrahim A and Iqbal F 2018 *Toxicol. Rep.* **5** 765 https://doi.org/10.1016/j.toxrep.2018.06.006

[3] Nikolova M P, Yankov E, Hikov T, Dechev D, Ivanov N, Parchorov S, Zaharieva V and Petrov P 2018 *IOP Conf. Series: Mater Sci & Eng.* **416** 012058 doi:10.1088/1757-899X/416/1/012058

[4] Nikolova M P, Nikolova V, Ivanova V L, Valkov S, Petrov P and Apostolova M D 2020. *Mater.Today Proc.* in press https://doi.org/10.1016/j.matpr.2020.05.051

[5] Lai Y-S, Chen W-C, Huang C-H, Cheng C-K, Chan K-K and Chang T-K 2015 *PLoS ONE* **10** e0127293 DOI: 10.1371/journal.pone.0127293

[6] Hainsworth S V and Soh W C 2003 *Surf. Coat. Technol.* **163–164** 515 https://doi.org/10.1016/S0257-8972(02)00652-7

[7] Shi L, Shi L, Wang L, Duan Y, Lei W, Wang Z, Li J, Fan X, Li X, Li S and Guo Z 2013 *Plos ONE* **8** e55015 doi: 10.1371/journal.pone.0055015

[8] Nikolova M, Nikolov V, Valkov S, Yankov E, Zaharieva V and Petrov P 2019 *Key Engineering Materials* **813** 178 https://doi.org/10.4028/www.scientific.net/KEM.813.178

[9] Maksakova O, Simoẽs S, Pogrebnjak A, Bondar O, Kravchenko Y, Beresnev V and Erdybaeva N 2018 *Mater Charact.* **140** 189 DOI: 10.1016/j.matchar.2018.03.048

[10] Bait L, Azzouz L, Madaoui N and Saoula N 2017 *Appl. Surf. Sci.* **395** 72 https://doi.org/10.1016/j.apsusc.2016.07.101

IOP Conf. Series: Materials Science and Engineering 1056 (2021) 012008 doi:10.1088/1757-899X/1056/1/012008

Surface modification of Co-Cr-Mo alloys by electron-beam treatment

S Valkov[1], S Parshorov[2], A Andreeva[3], M Nikolova[4] and P Petrov[1]

[1] Acad. E. Djakov Institute of Electronics, Bulgarian Academy of Sciences, 72 Tsarigradsko Chaussee, 1784 Sofia, Bulgaria
[2] Acad. A. Balevski Institute of Metal Science, Equipment and Technology with Hydro and Aerodynamic Center, Bulgarian Academy of Sciences, 67 Shipchenski Prohod blvd., 1574 Sofia, Bulgaria
[3] Faculty of Physics, St Kliment Ohridski University of Sofia, 5 James Bourchier blvd., 1164 Sofia, Bulgaria
[4] University of Ruse, Department of Material Science and Technology, 8 Studentska Str., 7017 Ruse, Bulgaria

E-mail: stsvalkov@gmail.com

Abstract. We present a study on the possibility of modifying the structure and properties of Co-Cr-Mo alloys by using the electron-beam surface modification technique. The technological parameters were optimized in order to obtain the highest cooling rate without melting the surface. The samples were characterized in terms of their phase composition and microstructure by X-ray diffraction (XRD) and scanning electron microscopy (SEM), respectively. The surface roughness was evaluated by atomic force microscopy (AFM). The corrosion properties were investigated by electrochemical tests. It was found that the treatment process leads to a transition in the phase composition, from $\varepsilon + \gamma$ to a single-phase γ structure. The electron-beam surface modification process causes an increase in the surface roughness and a significant improvement of the corrosion resistance.

1. Introduction

Currently, the Co-Cr-Mo alloy is an attractive material for applications in the modern medicine due to its biocompatibility and suitable mechanical properties and is widely used for manufacturing of joint replacements [1-3]. However, the biological corrosion leads to a release of metallic ions, which in turn leads to adverse reactions and failure of the implant [4]. Since the surface of the material is exposed during the implant's service, the discussed limitations can be overcome by an appropriate technology of surface treatment.

Recently, surface treatment of materials by high-energy fluxes (e.g., laser or electron beams) has become a common approach to improving the surface properties of materials due to the possibility of controlling precisely the input energy, the very short process time necessary, etc. The mechanism behind the electron-beam technologies is the transformation of the electrons' kinetic energy into heat bringing about a specific heat distribution from the surface to the bulk. The heating and cooling rates are quite high, which leads to changes in the phase and chemical composition, the surface topography, the microstructure, etc. [5-8]. Wei et al. [9] have investigated the possibilities of optimizing the hardness and tribological performance of a Co-Cr-Mo implant alloy by means of laser interference lithography; the results demonstrated a significant enhancement in the discussed properties.

However, the improvement of the surface properties of Co-Cr-Mo alloys by means of electron-beam treatment (EBT) has not yet been extensively studied. Furthermore, analyses of the EBT process effect on the surface topography and corrosion properties of Co-Cr-Mo-based alloys are currently lacking in the scientific literature. The study reported here aims to reveal the influence of electron-beam surface treatment of Co-Cr-Mo-based alloy on the material's surface topography and corrosion resistance. These properties are of major importance for the practical application of such implants; the results are expected to add knowledge to and extend the understanding of the processes of manufacturing modern implant materials.

2. Experimental

The Co-Cr-Mo alloy used had a chemical composition of 25.5 wt % Cr, 0.2 wt % Mn, 4.3 wt % Fe, 59.9 wt % Co, 1.0 wt % Ni, 4.7 wt % Mo, 0.6 wt % W and 3.8 wt % other elements. The surfaces of specimens with dimensions of 20×20 mm and a thickness of 4 mm were electron-beam treated by Leybold Heraeus (EWS 300/15-60) electron-beam equipment using a continuous electron-beam scanning. The technological parameters were as follows: accelerating voltage 52 kV, electron-beam current 25 mA, electron-beam scanning frequency 1 kHz, speed of specimen motion 40 mm/sec; electron-beam diameter 0.5 mm, linear electron-beam trajectory. These technological parameters were optimized in order to obtain a treated zone without the formation of a melt pool. However, partial melting of some parts of the surface of the sample probably takes place during the scanning electron-beam treatment process. The temperature was evaluated at about 1200 ^{0} C by using a numerical model published in works [10, 11].

The samples were characterized by X-ray diffraction (XRD) with CuKα (1.54 Å) radiation. The experiments were carried out in a symmetrical Bragg-Brentano (B-B) mode at a 2θ scale within the range from 20° to 100° with a step of 0.03° and a counting time of 5 s per step.

The samples were further studied by means of a LYRA 3 XMU scanning electron microscope (SEM, Tescan), equipped with an EDX detector (Quantax 200, Brucker), which integrates a true standardless analysis with P/B ZAF corrections (Z is the atomic number correction factor; A, the X-ray absorption correction factor; F, the fluorescence correction factor). During the experiments, back-scattered electrons were used.

The surface topography of the specimens was studied by atomic force microscopy (AFM). The force constant was 43 N/m and the radius of the silicon nitride tip was 10 nm. The experiments were performed in a non-contact mode.

The experiments on determining the corrosion properties of the samples were conducted in an electrochemical cell with a platinum counter electrode and a saturated Ag/AgCl reference electrode. Both electrodes together with the working one with a 0.2-cm^2 exposed area were immersed in 80 ml of naturally aerated artificial saliva solution (Fusayama) containing KCl (0.4 g/l), NaCl (0.4 g/l), $CaCl_2{\cdot}2H_2O$ (0.795 g/l), $NaH_2PO_4{\cdot}2H_2O$ (0.690 g/l), $Na_2S{\cdot}9H_2O$ (0.005 g/l), CH_4N_2O (1 g/l) with pH 5 (decreased with 1 N HCl) at 37±0.05 °C. After OPC carried out for 30 min, potentiodynamic polarization curves were recorded by sweeping the potential starting at a cathodic potential of about 250 mV below the OPC up to +2000 mV *vs.* Ag/AgCl at a scan rate of 1 mV s^{-1} using a potentiostat 263A (EG&G Princeton Applied Research) coupled to a PC by a controller. The corrosion parameters, E_{corr}, (corrosion potential) and I_{corr} (corrosion current density) were obtained from the polarization curves by using the Tafel extrapolation method. The polarization resistance, R_p, was determined from the Stern-Geary equation.

3. Results and discussion

Fig. 1 presents the XRD results. The diffraction pattern in Fig. 1 (a) pertains to the initial Co-Cr-Mo alloy, while Fig. 1 (b)) is the diffractogram of the electron-beam modified specimen. The pattern of the untreated sample exhibits some peaks of ε-Co and, in addition, there is one diffraction maximum at about 2θ of 44.45° belonging to γ-Co. It should be noted that the ε-Co phase has a hexagonal closed-packed (hcp) crystal lattice which is stable at room temperature and transforms to γ at 900 °C. The latter phase has a face-centered cubic (fcc) structure. The phase transition fcc-to-hcp during the alloy manufacturing usually occurs slowly and some traces of the metastable at room temperature fcc phase remain [12, 13]. During the electron-beam surface treatment of the Co-Cr-Mo alloy, (Fig. 1 (b)), a phase transition takes place to the fcc-γ phase. During the EBT process, the surface of the sample is heated up to a certain temperature, where the gamma phase is stable. As already mentioned, the transformation fcc-to-hcp is too slow and cannot occur during the following rapid cooling process. It should be noted that the results obtained by the XRD analysis are in agreement with those published in Ref. [9] where a surface modification of Co-Cr-Mo implant alloy by laser interference lithography has been studied.

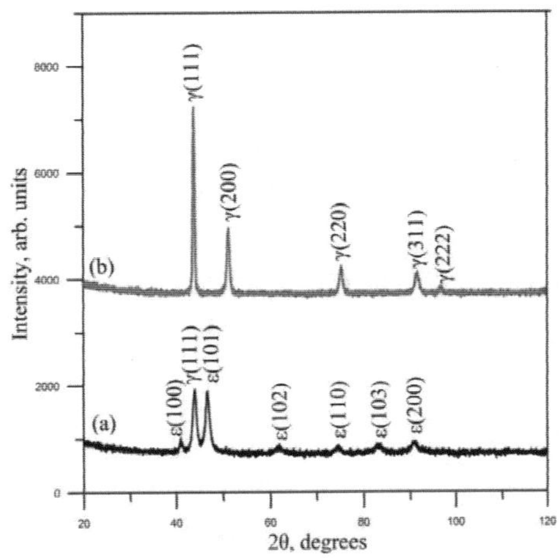

Figure 1. X-ray diffraction pattern of the Co-Cr-Mo alloys a) untreated; b) electron-beam treated.

Figure 2 presents a scanning electron microscopy (SEM) image of the untreated Co-Cr-Mo sample (Fig. 2 (a)). The chemical composition of the alloy was studied by EDX and the results are shown in Fig. 2 (b) and Table 1. One can see the biphasic ε+γ microstructure of the discussed specimen. The main phase is ε surrounded by the line-like regions of γ phase. The SEM results are in agreement with those of the XRD analysis pointing to a presence of the discussed bimodal structure.

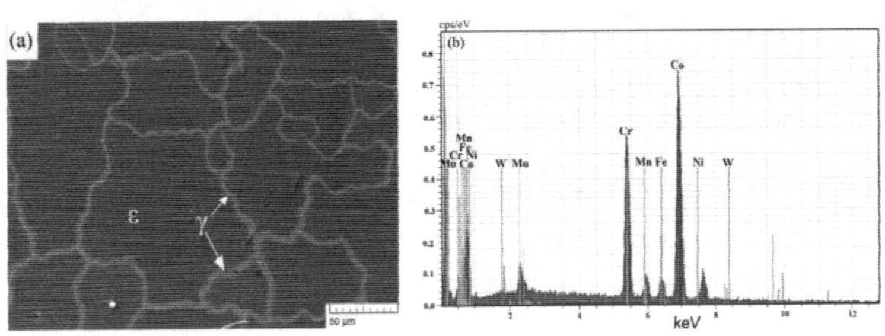

Figure 2. a) SEM image of the untreated Co-Cr-Mo sample, and b) the corresponding EDX spectrum.

Figure 3 presents a SEM micrograph of the treated Co-Cr-Mo sample (Fig. 3 (a)). The results of the EDX study of this sample's chemical composition are shown in Fig. 3 (b) and Table 1. After the EBT, the microstructure becomes a single-phase structure and no traces of the second phase can be seen. These results match those of the XRD experiments, where a single γ-Co phase was registered.

Therefore, the phase transformation from ε-Co + γ-Co to γ-Co after the electron-beam surface modification of the Co-Cr-Mo alloy observed by the XRD experiments is confirmed by the SEM analysis. The EDX measurements, presented in Figs. 2 (b) and 3 (b), were conducted in order to evaluate the influence of the electron-beam surface modification on the chemical composition of the Co-Cr-Mo alloys; the results are summarized in Table 1. It is evident that the surface treatment leads to an increase in the amount of Cr and a decrease in the content of Co attributed to the evaporation of cobalt due to its lower melting temperature.

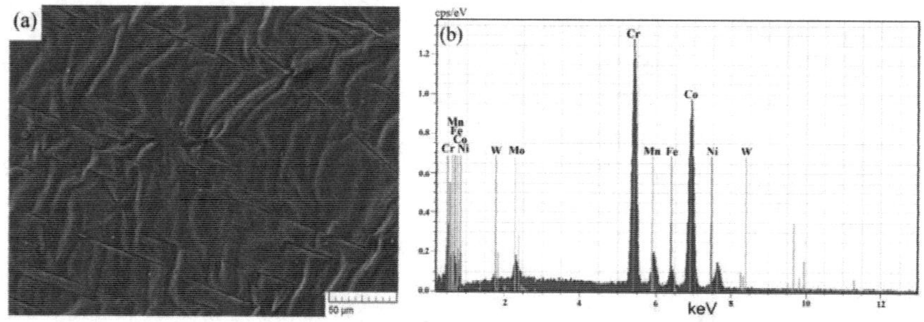

Figure 3. a) SEM image of the treated Co-Cr-Mo sample, and b) the corresponding EDX spectrum.

Table 1. Chemical composition of the Co-Cr-Mo alloy with and without electron-beam surface modification

	Cr	Mn	Fe	Co	Ni	Mo	W	Other
Untreated alloy	22.2	0.3	4.5	59.6	1.6	5.1	1.0	5.7
Treated alloy	34.3	1.5	3.4	49.1	1.1	3.1	3.6	3.9

Fig. 4 shows three-dimensional AFM images of the specimens' surface topography. It is obvious that the EBT process strongly affects the surface topography.

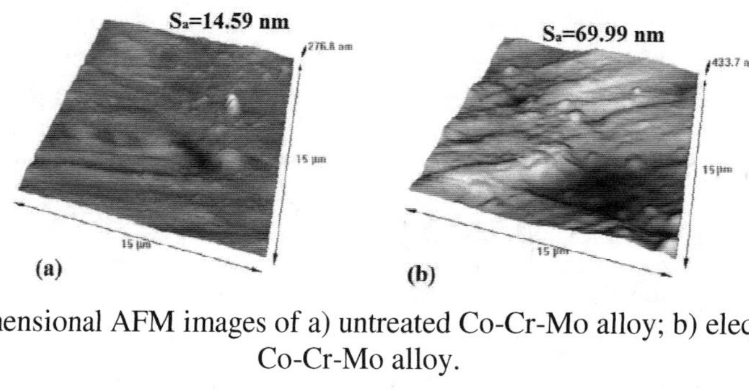

Figure 4. Three dimensional AFM images of a) untreated Co-Cr-Mo alloy; b) electron-beam treated Co-Cr-Mo alloy.

The surface roughness was quantitatively evaluated by mean of the roughness parameter S_a, which is given by

$$S_a = \frac{1}{MN}\sum_{k=0}^{M-1}\sum_{l=0}^{N-1}|z(x_k, y_l) - \mu| \quad (1)$$

In (1), M and N are the numbers of measured points along the x and y axes; z is the height of each measured point with coordinates x_k and y_l; μ is the mean height. The results obtained for the parameter S_a are 14.59 nm for the untreated alloy and 69.99 nm for the EBT Co-Cr-Mo alloy. Thus, the electron-beam surface modification leads to an about fivefold increase in the surface roughness, which can be due to the evaporation of a small amount of the treated material during the EBT and the subsequent condensation of the vapors. According to the authors of [14], the vapors condensation can form protrusions thus increasing the surface roughness. It should be noted that the increase observed in the surface roughness after the EBT is associated with an increase in the contact surface area, which enhances the cell adhesion and supports the cell growth. Furthermore, according to the authors of [15, 16], such nano-rough surface architecture is less susceptible to being colonized by bacterial cells in comparison with the conventionally manufactured surfaces.

Fig. 5 shows experimental polarization curves of untreated and EBT Co-Cr-Mo alloy in artificial saliva (pH 5) at 37±0.05 °C. The results demonstrate a decrease in the corrosion current density and an increase in the corrosion potential values pointing to a significant improvement in the corrosion resistance of the treated sample compared with the untreated alloy. It is known that lower values of the corrosion current density correspond to a lower corrosion rate. At the same time, the increase in the E_{corr} indicates a better corrosion resistance of the tested material [17-19]. The change in the relative chemical elements content at the surface of an alloy alters the oxide resistance and, therefore, the corrosion rate. According to Dobrzański and Reimann [20], the Cr concentration plays a major role in the corrosion behavior, the higher chromium content corresponding to a better performance.

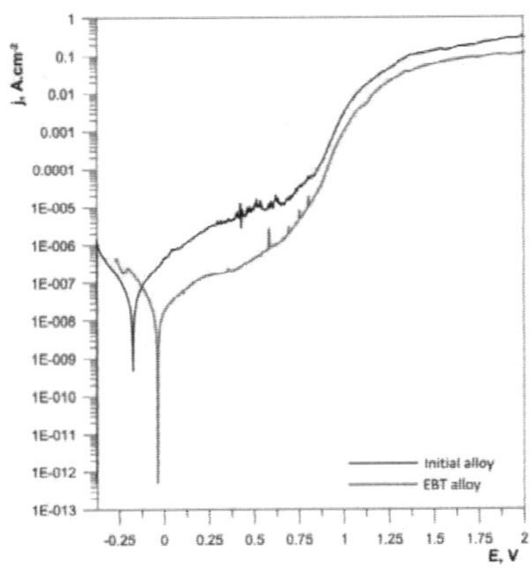

Figure 5. Potentiodynamic polarization curves of initial and treated samples in artificial saliva (pH 5) at 37±0.05 °C.

4. Conclusions

The results obtained in the present study show the possibility of modifying the structure and properties of Co-Cr-Mo alloys by means of electron-beam surface treatment. The treatment leads to a phase transformation, from main ε phase and a small amount of γ phase, to a single-phase γ structure due to the fast cooling rate. The depth of the treated zone is about 250 μm. It is also shown that the use of electron-beam surface modification results in an about fivefold increase in the surface roughness corresponding to an increase in the contact surface area. Furthermore, the results of the electrochemical experiments demonstrate a significant improvement in the corrosion resistance following the electron-beam surface modification of the Co-Cr-Mo alloy.

Acknowledgments: The work was financially supported in part by the Bulgarian National Science Fund under project No DN 17/17.

References

[1] Al Jabbari Y 2014 *J. Adv. Prosthodont.* **6** 138–145
[2] Vendittoli P, Amzica T, Roy A, Lusignan D, Girard J and Lavigne M 2011 *J. Arthroplasty.* **26** 282–288
[3] Niinomi M 2002 *Metall. Mater. Trans. A* **33** 477–486
[4] Abu-Amer Y, Darwech I and Clohisy J 2007 *Arthritis Res. Ther.* **9** S6

[5] Weglowski M, Blacha S and Phillips A 2016 *Vacuum* **130** 72-92
[6] Ormanova M, Petrov P and Kovacheva D 2017 *Vacuum* **135** 7-12
[7] Petrov P, Dechev D, Ivanov N, Hikov T, Valkov S, Nikolova M, Yankov E, Parshorov S, Bezdushnyi R and Andreeva A 2018 *Vacuum* **154** 264-271
[8] Nevskii S, Sarychev V, Konovalov S, Granovskii A and Gromov V 2020 *Metals* **10** 1399
[9] Wei X, Li W, Liang B, Li B, Zhang J, Zhang L and Wang Z 2016 *Tribology International* **97** 212-217
[10] Angelov V, Ormanova M, Kaisheva D, Lazarova R, Dimitrova R and Petrov P 2019 *Nucl. Instr. Meth. Phys. Res. Sec. B* **440** 88–94
[11] Ormanova M, Angelov V and Petrov P 2016 *J. Phys. Conf. Ser.* **700** 012033
[12] Mani A, Salinas-Rodrigez A and Lopez H 2011 *Mater. Sci. Eng. A* **528** 3037–3043
[13] Wei D, Koizumi Y, Takashima T, Nagasako M and Chiba A, 2018 *Mater. Res. Lett.* **6** 93-99
[14] Zhang K, Zou J, Grosdidier T, Gey N, Weber S, Yang D and Dong C 2007 J. Vac. Sci. Technol. **25 (1)** 28–36
[15] Curto B, Brunella M, Giordano C, Pedeferri M, Valtulina V, Visai L and Cigada A 2005 *Int. J. Artif. Organs* **28 (7)** 718–730
[16] Pokrowiecki R, Szaraniec B, Chłopek J and Zaleska M 2014 *Eng. Biomater.* **124** 2–10
[17] Liu C, Zhou Z and Li K 2017 *Electrochemica Acta* **241** 331-340
[18] Jin W, Wu G, Feng H, Wang W, Zhang X and Chu P 2015 *Corros. Sci.* **94** 142–155
[19] Zhang T, Deng Q, Liu B, Wu B, Jing F, Leng Y and Huang N 2015 *Surf. Coat. Technol.* **273** 12–19
[20] Dobrzański L and Reimann L 2011 *J. Achieve. Mater. Manuf. Eng.* **49** 193–199

Biogenic ferroxides derived from *Leptothrix* bacteria for applications in electronics, biomedicine and biotechnology

R Angelova[1], L Slavov[1], B Blagoev[2], Ch Ghelev[1], M Iliev[3], V Groudeva[3] and I Nedkov[1]

[1]Institute of Electronics, Bulgarian Academy of Sciences, Sofia, Bulgaria
[2]Institute of Solid State Physics, Bulgarian Academy of Sciences, Sofia, Bulgaria
[3]Faculty of Biology, St. Kliment Ohridski University of Sofia, Sofia, Bulgaria

E-mail: raly_angelova@abv.bg

Abstract. The present work is focused on studying by-products derived from the metabolism of bacteria of the *Leptothrix* genus, which are among the first described microorganisms associated with the iron cycle in nature. The products of their metabolism are nanostructured biogenic iron oxides in the form of precipitating powders and sheath structures. The sheath structures can be considered as an organic matrix in which inorganic crystallites are discretely dispersed. We used X-ray diffraction, magnetic measurements, light microscopy and scanning electron microscopy to characterize biogenic products formed in a silicon iron glucose peptone medium under laboratory conditions. The studies showed a lack of significant differences between the naturally obtained and the artificially synthesized biogenic sheaths, i.e., an adequate laboratory technological process had been developed. From the point of view of nanoelectronics application, these biogenic by-products are unique because they are biocompatible, have specific electromagnetic properties and are potential candidates for various applications in biomedicine and electronics.

1. Introduction

The bacteria from the *Leptothrix* genus are typical β-proteobacteria capable of oxidizing Fe^{2+} and Mn^{2+} at neutral pH and are commonly found in aquatic environments [1, 2]. They grow as chains of cells into filaments 0.4 –7 μm in width [1].

This work is focused on studying by-products derived from the metabolism of *Leptothrix* bacteria. These bacteria do not pose a danger to the human body. Products of their metabolism are nanostructured biogenic iron hydroxides in the form of yellow/brown mats, which include also their sheath structures. As a result of the oxidation of iron cations ($Fe^{2+} \rightarrow Fe^{3+}$), different iron containing compounds are produced (oxyhydroxides, such as lepidocrocite, γ-FeOOH, etc.). The biotic or abiotic processes can continue in the walls of the sheath structures until ferroxide structures, such as magnetite (Fe_3O_4), maghemite (γ-Fe_2O_3), are formed. The review of the literature shows that there exist unclear points regarding the causes of oxidation of iron ions and their contribution to the metabolism of these environmentally important bacteria. The results of the research related to the formation of sheath structures are quite contradictory, as are the factors that influence their formation [1]. The sheath structures can be considered as an organic matrix in which inorganic crystallites are discretely dispersed. From the point of view of nanoelectronics application, these biogenic by-products are unique because they are biocompatible, have specific electromagnetic properties and can be potential candidates for various applications in biomedicine, electronics [2], as pigments, catalysts, adsorbents, carriers [3-8] and as precursors for synthesis of the hybrid battery–super capacitor system „bioFe_2O_3+AC)//$LiBF_4$//AC“ [6]. Magnetized sheath structures could be oriented in a magnetic field higher than the saturation magnetization of the impregnated sheath structures; together with their anisotropic properties; this makes them suitable for creating biochips for use in electronics. These bioproducts open possibilities for application in biomedicine, for example radio frequency protection of pacemakers or other implants in a living organism.

The interest in materials produced by *Leptothrix* bacteria has grown in recent years [7, 9-12]. We present a study on iron bio-ferroxides in the sheathed structures of *Leptothrix* sp. during laboratory cultivation in a selective medium (SIGP medium [2, 10]). Such thorough investigations could have a major impact on applications concerning the industrial synthesis of Fe(III)-bearing compounds and on the possibility for these techniques to be replaced by environmentally-friendly biosynthesis procedures [13].

2. Materials and Methods

A pure culture from *Leptothrix* genus was used isolated from a natural stream in Vitosha Mountain located at an altitude of 1783 m (42°35’15”N/23°14’55”E) [10]. For the formation of sheath structures, laboratory cultivation was performed in a silicon iron glucose peptone (SIGP) medium [10]. The isolate was inoculated into the medium with iron cuttings and glass plates on the bottom serving for attachment of the bacteria and cultivated under dynamic conditions (70 rpm) at 20 °C.

Studying the morphology of the by-products formed in an SIGP medium under laboratory conditions was performed by light microscopy and scanning electron microscopy (SEM) (JEOL JSN-5510, JEOL, Japan). The structural characterization of the biogenic iron oxides was carried out by X-ray diffraction using a Bruker D8 diffractometer in the Bragg–Brentano reflection geometry with Cu Kα radiation (λ = 1:5418 Å). The magnetic measurements were measured by a physical properties measurement system (PPMS, Quantum Design).

3. Results and Discussion

The cultivation of *Leptothrix* sp. in an SIGP medium resulted in an increase in the fluffy ocherous deposits in the flasks. Light micrograph (figure 1a) and SEM (figure 1b) images revealed the approximate size of the sheath structures formed in a seven-day period of cultivation – around 10 μm with an average diameter of up to 1 μm (figures 1 and 2). Their diameters and lengths varied, probably depending on their maturity [2]. Spring [11] reported that the sheaths structures of *Leptothrix* sp. are brittle and easily broken into shorter fragments. Therefore, measurements of the length may not be reliable for these structures.

a) b)

Figure 1. Light a) and SEM b) images of sheath structures formed in an SIGP medium.

The sheath structures obtained under laboratory conditions were compared with those taken from the streams of Vitosha Mountain using light microscopy and SEM (figure 2). Such structures are found in the natural environment samples; most of the sheaths are empty – they do not contain bacteria and their structure fully corresponds to the descriptions given by other authors [2, 12].

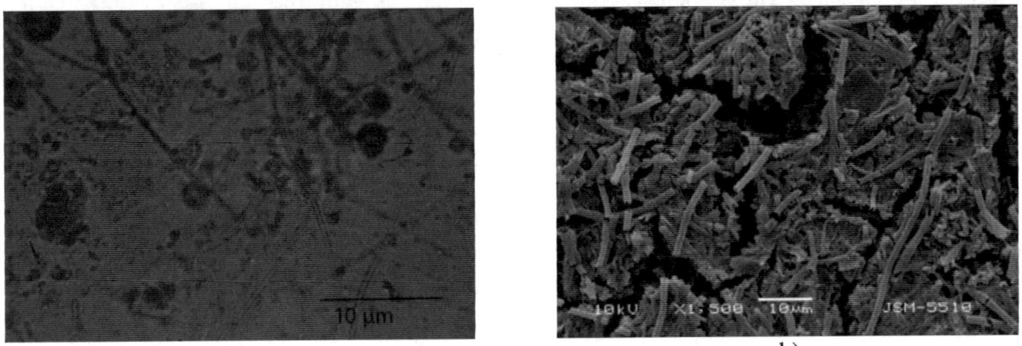

a) b)

Figure 2. Light a) and SEM b) images of sheath structures grown in a natural environment.

The studies showed a lack of significant differences between the naturally and laboratory obtained biogenic sheaths, i.e., an adequate laboratory technological process had been developed and used for the production of such a useful material.

The results about the dynamics of the ferrous ions concentration during cultivation of the bacteria in an SIGP medium measured by titration with potassium permanganate (0,1 N $KMnO_4$) are given in figure 3. The rate of chemical oxidation of the ferrous ions is very low in the control sample tested.

The XRD data from our previous investigation of the same sample showed a single-phase composition of the iron-containing particles on the sheaths structures obtained (lepidocrocite – γ-FeO(OH)), as well as a uniform size of the iron oxihydroxide particles – 8 nm [10]. The spectrum confirmed the presence of iron oxide compounds in the walls of the sheath structures (the spectrum was obtained from a coating of biogenic sheath structures on a glass substrate and shows also the background lines of the substrate). Lepidocrocite is antiferromagnetic and in practice the sheath structures have no magnetic properties at room temperature in the earth's magnetic field.

The SEM/EDX analysis in our previous investigation [10] revealed that S, O, C, K, Ca, Na, Si, Fe, and P are the inorganic elements of cultured mature sheath structures. These results agree with those of other studies, which reported that the major inorganic component of the sheath structures is in the form of iron oxides that bind other inorganics, such as Si, P, and often Ca and S [14, 15].

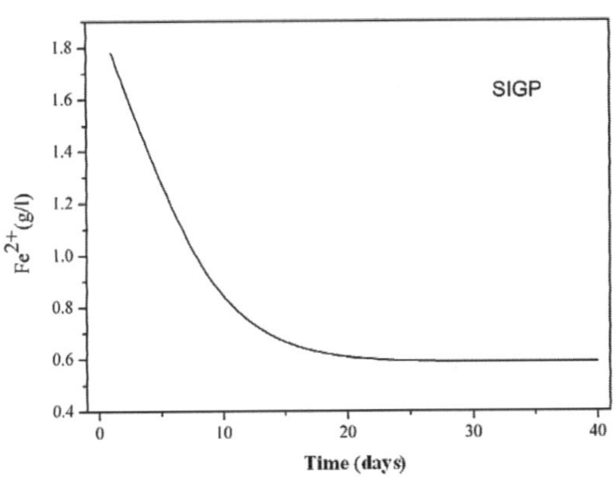

Figure 3. Dynamics of oxidation of the ferrous ions during cultivation in an SIGP medium.

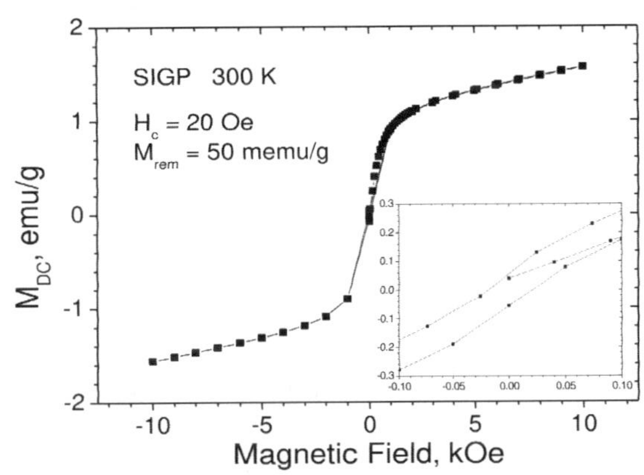

Figure 4. Magnetic hysteresis curve $M(H)$ at 300 K of an SIGP sample.

The magnetic measurements performed by a physical property measurement system (PPMS) of the SIGP sample showed a weak magnetic behavior ($M \sim 1.6$ emu/g at 10 kOe) (figure 4).

The by-product obtained from the SIGP medium exhibited the properties of a soft magnetic material ($H_c \sim 20$ Oe, $M_{rem} \sim 50$ memu/g) with an indication of the presence of nanoscale particles at room temperature, which we associate with structural changes in the surface of the discretely dispersed lepidocrocite particles in the organic matrix of the sheath structures.

Such magnetic behavior is expected to arise from a relatively small quantity of low-magnetic-response material dispersed in a non-magnetic matrix.

The results of the magnetic measurements at room temperature of a powder sample from the SIGP medium with a high content of biogenic sheath structures showed nearly paramagnetic properties. Such sheath structures of biogenic origin can be enriched with nanoscale magnetite by a simple technological procedure (co-precipitation at room temperature), which can enhance the magnetic response of the material, while keeping the magnetic structure intact. Sheathed structures enriched with such magnetite nanoparticles can be capable of adsorbing various substances that endanger the environment, i.e., dyes, Fe^{2+} ions, heavy metals [16, 17].

4. Conclusion

The insoluble hydroxides formed by *Leptothrix* bacteria after oxidation of the ferrous iron are deposited on the bacterial surface, and specific sheath structures are formed after labortary cultivation in an SIGP medium. The iron-containing part of the sheaths is single-phase lepidocrocite exhibiting a weak magnetic behavior at room temperature. These sheath structures, as well as the hydroxide, are of great interest for application in various nanotechnologies, biomedicine and bioengineering as ferrofluids, pigments, adsorbents, or as a contrast agent for magnetic resonance imaging, detoxification of biological fluids and many others.

Acknowledgements

This work was supported by the Bulgarian National Science Fund of the Ministry of Education and Science under project T02-17/2014.

References

[1] Cornell R M and Schwertmann U 2003 *"Iron Oxides"* 2nd edn (John Wiley & Sons, New York) p. 484

[2] Sawayama M, Suzuki T, Hashimoto H, Kasai T, Furutani M, Miyata N, Kunoh H and Takada J 2011 *Curr. Microbiol.* **63** 173

[3] Gupta A K and Gupta M 2005 *Biomaterials* **26** 3995

[4] Kunoh T, Kunoh H and Takada J 2015 *J. Microb. Biochem. Technol.* **7** 419

[5] Kunoh T, Hashimoto H, Suzuki T, Hayashi N, Tamura K, Takano M, Kunoh H and Takada J 2016 *Minerals* **6** Article No. 4

[6] Veleva S, Angelova R, Stoyanov L, Groudeva V, Kovacheva D, Mladenov M and Raicheff R 2014 Nanosci. *Nanotechnol.* **14** 50

[7] Safarik I, Angelova R, Baldikova E, Pospiskova K and Safarikova M 2017 *Mater. Sci. Eng. C* **71** 1342

[8] Shopska M, Cherkezova-Zheleva Z, Paneva D, Iliev M, Kadinov G, Mitov I and Groudeva V 2013 *Open Chemistry* **11** 215

[9] Angelova R, Iliev M, Mitova M, Voynova V, Tasheva L, Slavov L, Groudeva V and Antonova-Nikolova, S 2014 *J. BioSci. Biotech.* 67-70

[10] Angelova R, Iliev M, Slavov L, Koutzarova T, Closset R, Nedkov I and Groudeva V 2015 *J. BioSci. Biotech.* 165-170

[11] Spring S 2006 *"The genera Leptothrix and Sphaerotilus" in Prokaryotes* vol 5 ed Dworkin M, Falkow S, Rosenberg E, Schleifer K-H, Stackebrandt E (Springer, New York) pp 758–777

[12] Ellis D 2003 *"Microbiology of the Iron-Depositing Bacteria"* (Wexford College Press, Palm

Springs)
[13] Mosiniewicz-Szablewska E, Safarikova M and Safarik I 2010 *J. Nanosci. Nanotechnol.* **10** 2531
[14] Sakai T, Miyazaki Y, Murakami A, Sakamoto N, Ema T, Hashimoto H, Furutani M, Nakanishi M, Fujii T and Takada J 2010 *Org. Biomol. Chem.* **8** 336
[15] Suzuki T, Hashimoto H, Matsumoto N, Furutani M, Kunoh H and Takada J 2011 *Appl. Environ. Microbiol.* **77** 2877
[16] Safarik I, Filip J, Horska K, Nowakova M, Tucek J, Safarikova M, Hashimoto H, Takada J and Zboril R 2015 *Sci. Technol.* **12** 673
[17] Rentz J A, Turner I P and Ullman J L 2009 *Water Res.* **43** 2029-2035

IOP Conf. Series: Materials Science and Engineering 1056 (2021) 012010 doi:10.1088/1757-899X/1056/1/012010

Investigation of the deviation during the information transfer from the prosthetic field to the laboratory scanners

H. Galeva[1]*, T. Uzunov[1], Y. Stoev[1], Y. Sofronov[2], G. Todorov[2]

[1] Faculty of Dental Medicine, Medical University, Sofia, Bulgaria
[2] Technical University of Sofia, 8 St. Kliment Ohridski blvd., Sofia, Bulgaria

*e-mail:dr.hrgaleva@gmail.com

Abstract. The aim of this work is to investigate the deviations during the first stage of the information transfer from the prosthetic field to the dental laboratory scanners for CAD/CAM technology. The first upper left molar on an artificial acrylic resin dental model was prepared for full metal-ceramic crown according to the standard guidelines for step cut. The prepared prosthetic field was scanned by a high-precision industrial scanner with a 1-mm ruby ball stylus and a digital copy was made. With a standard tray, a two-phase two-stage silicon (A-silicone Zhermack hydrorise putty and light body) impression from the acrylic model was taken following the technology recommended by the manufacturer. The impression was powdered with dental stone type IV and a stone model was created 1 h after impression, the time delay taking place ordinarily in dental practice. Using the same measurement methods, the stone model's surfaces were digitized. Both digital models were compared using the acrylic resin dental model as a reference. The results on the deviation during the transfer of information between the prosthetic field and the stone model represent the first level of inaccuracy in the CAD/CAM processing technology. The difference in the alteration of each dimension according to the reference and the stone model is not more than 20 μm in shrinkage or expansion direction. Thus, the widely used approach for information transfer using stone models from the prosthetic field to laboratory scanner for CAD/CAM technology could be considered as relevant with an accuracy of about ±10 μm. If the silicon impression is performed carefully, the stone models are sufficiently accurate representations of the prosthetic field and their digitalization in the dental laboratory is a reliable starting point for the CAD/CAM technology.

1. Introduction

Creating a precise prosthetic construction requires a correct and accurate reproduction of the prosthetic field in the dental laboratory. Taking a correct impression and casting a suitable stone model is a guarantee that the future construction will fit well.

Among the conditions that the impression must meet is avoiding any volumetric changes and reproducing in detail the prosthetic field [1].Usually, the impression is powdered with gypsum to create a master model. However, gypsum has some expansion that must be taken into account when choosing an impression material [2].

CAD/CAM systems have been increasingly used for a number of reasons, such as being very precise and saving time for both patients and doctors [3]. Various possibilities exist of transferring the information from the prosthetic field to the laboratory. Scanning the dentition by an intraoral scanner or scanning physical plaster models by a desktop optical scanner are some of them [4]. According to

IOP Conf. Series: Materials Science and Engineering 1056 (2021) 012010 doi:10.1088/1757-899X/1056/1/012010

Nisa Gül Amuk and al., a digital model could be an alternative to the plaster model [5]. A study conducted by Jin-Hun Jeon and al. showed that there is a significant difference between the accuracy achieved by the scanning impression and the stone model [6]. Some authors, as Jan-Frederik Güth and al., found that the digital impressions obtained by scanning a plaster model are less accurate that those obtained by impression scanning or direct scanning of the prosthetic field [7]. However, the correct information transfer is very important in view of the accuracy of the final construction. This is why the deviation during this transfer should be thoroughly investigated.

2. Method and material

For the purpose of this study, the plastic upper jaw model (SpofaDental) shown on fig. 1 was employed as a prosthetic field. Using a standard full-arch tray and additional silicon (Zhermack hydrorise putty and light body), a two layer impression was taken from the plastic model (fig. 2).

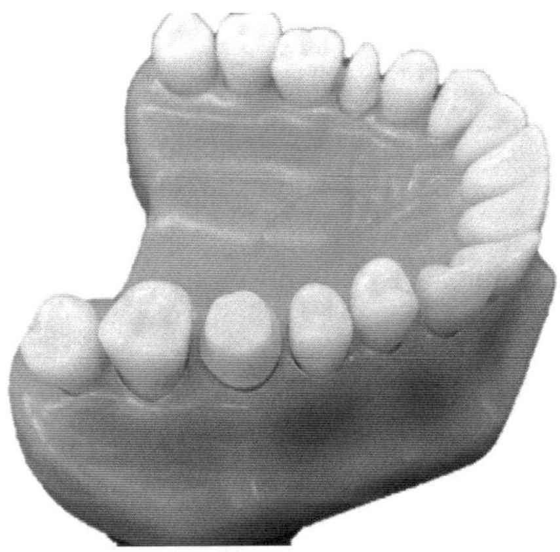

Figure 1. Upper jaw plastic model

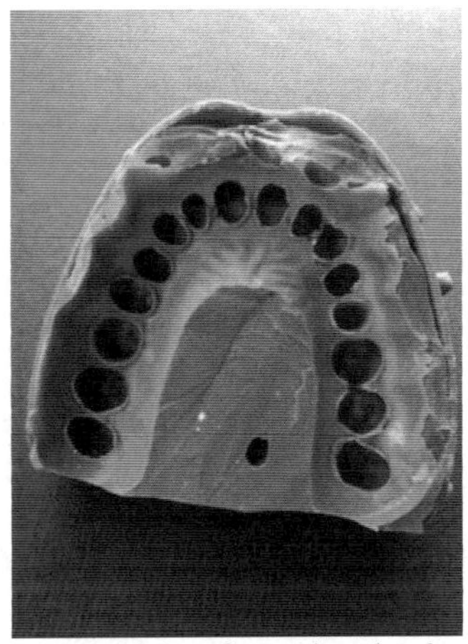

Figure 2. Impression from the plastic model

One hour later, the same impression was powdered with Gypsum class 4 (Elite rock), mixed according to the manufacturer`s instructions (fig. 3).

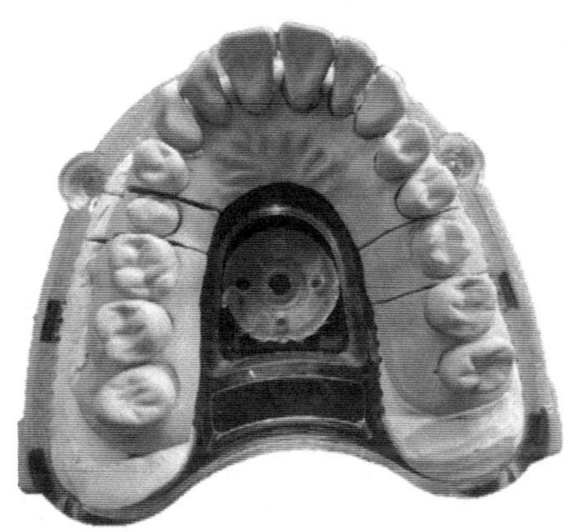

Figure 3. Stone model from the impression

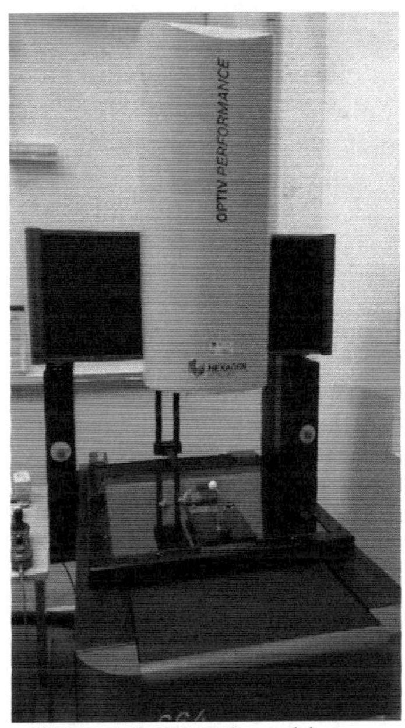

Figure 4. CMM machine

The plastic and the stone models were scanned by a coordinate measuring machine (Optiv Performance 664, Hexagon Metrology). This device combines optical and tactile sensors; we made use of a six-way touch-trigger probe with a 1-mm diameter ruby-ball stylus with an accuracy according to calibration of ±1.00 μm. Two clouds of points using the same strategy were obtained. The differences between them represent the deviation during the information transfer from the prosthetic field to the laboratory scanner.

3. Results

From each model, 1786 points were measured. Based on them, two 3D mesh models were produced from the cloud of points using the 3D Reshaper software, as shown in fig. 5. Then, the two digital models were superimposed and the difference between them were detected using Geomagic Control X software with the plastic one as a reference, fig. 6.
The comparative analysis showed that about 87.3% of measured points differ within ±100 μm, and 12.2% deviate by about ±10 μm and less. The deviation directions show that the stone model is uniformly shrunk compared with the reference model.

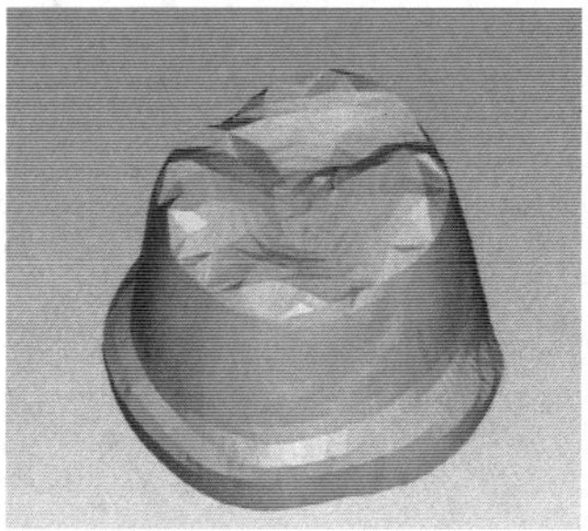

Figure 5. Reference 3D model

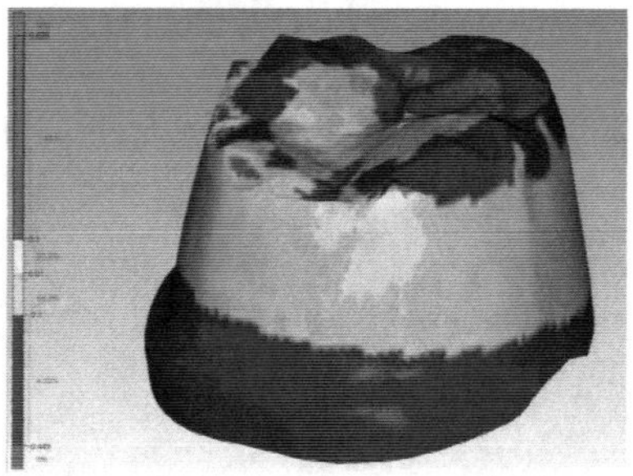

Figure 6. Comparative analysis

4. Discussion

The result achieved of the comparative analysis leads to several important conclusions. The first is that an error was made in the process of creating the stone model or in the measurements. But the process of creating the impression and the stone model was performed according to the manufacturer's recommendations and no deviations were visually observed before the measurements. The second is that the comparisons are based on a small number of measured points compared to the published in the literature comparative analyses made with non-contact measuring devices. This effect is avoided by comparing 3D mesh models and high measurement accuracy unattainable for contactless scanners. The third possibility is that in some cases this is the achievable accuracy of the process and the good fit is ensured by adjusting the stone model by a dental technician and then by the dentist.

5. Conclusions

The comparative study was performed based on 1786 points measured with high accuracy from reference acrylic and stone models.

The comparative analysis showed that the stone model is smaller by about ± 0.1 mm. This is a relatively large deviation and leads to contradictory conclusions. One of them is that the results are relevant to some cases where the stone model as a transfer of information in the initial stage of CAD/CAM technology will compromise the subsequent work of dental technicians. Such a deviation in the transfer of information could not be detected at an early stage by either the dentist or the dental technician. Further investigation will be conducted with the aim of clarifying the possible reasons of these results.

Acknowledgments: This research was funded by Medical University-Sofia under Grant 112/24.06.2020.

References

[1] Filchev A and Ralev R 2015 Preclinic of the prosthetic dentistry (Sofia) pp 171-173
[2] Anastasov I 2013 Dental materials (Sofia: Sofia smile center) pp 65-76
[3] Lee S J and Gallucci G O 2013 Digital vs. conventional implant impressions: efficiency outcomes Clin.Oral.Impl.Res. 24 111-115
[4] Taneva I and Uzunov T 2020 Influence of post polymerization processing over the mechanical features of 3d printed occlusal splints J. Phys.: Conf. Ser. 1492 012018
[5] Amuk N Karsli E and Kurt G 2019 Comparison of dental measurements between conventional plaster models, digital models obtained by impression scanning and plaster model scanning Int Orthodontics 151-158
[6] Jeon J Hwang S Kim J and Kim W 2018 Trueness and precision of scanning abutment impressions and stone models according to dental CAD/CAM evaluation standards J Adv Prosthodont 335-9
[7] Güth J Keul C Stimmelmayr M Beuer F and Daniel Edelhoff 2013 Accuracy of digital models obtained by direct and indirect data capturing Clin Oral Invest 1201–1208

IOP Conf. Series: Materials Science and Engineering 1056 (2021) 012011 doi:10.1088/1757-899X/1056/1/012011

Structural and mechanical properties of Ta-based coatings deposited on polymers for biomedical applications

N Donkov[1], A Zykova[2,3], V Safonov[2,3], S Dudin[3], S Yakovin[3], P Petrov[1], S Rabadzhiyska[1]
[1]E. Djakov Institute of Electronics, Bulgarian Academy of Sciences, 72 Tsarigradsko Chaussee, 1784 Sofia, Bulgaria
[2]National Science Center "Kharkov Institute of Physics and Technology",
1 Academicheskaya Str., 61108 Kharkiv, Ukraine
[3]V.N. Karazin Kharkiv National University, 4 Svobody Sq., 61022 Kharkiv, Ukraine

E-mail: nikolaidd@abv.bg

Abstract. The polypropylene meshes currently used for hernioplasty induce severe chronic inflammatory reactions over a long-term period of use. These processes lead to scarring and have a negative effect on the connective tissue and the abdominal walls. It is well-known that the biocompatibility of polymer surfaces might be improved by surface modification. We report on the structure and composition of nanostructured Ta, Ta_2O_5 and TaON coatings deposited on polypropylene in correlation with their mechanical properties. The elemental distribution and chemical composition of the coatings were analyzed by energy dispersive X-ray spectroscopy. The coatings cross-sectional structure and morphology were observed by electron scanning microscopy. Mechanical tests were performed of the polypropylene and of the samples with Ta-based coatings to evaluate the Vickers hardness and the elastic modulus. The coated samples exhibited higher nano-hardness and elasticity values, significantly exceeding the values of untreated polypropylene samples. The Ta-based coatings deposited by magnetron sputtering improve the mechanical characteristics and biocompatibility of the polypropylene materials in view of, e.g., hernia repair.

1. Introduction

Ta-based materials and coatings are excellent endoprosthetic biomaterials for use in trauma and cardio-vascular surgery. Tantalum has also been selected as an alternative to other commonly used metallic material for orthopedic implant applications. Further, the biocompatibility and antibacterial properties of tantalum coatings have been reported. Tantalum-based thin films exhibited antimicrobial potential against several microorganisms, such as *Staphylococcus aureus*, *Staphylococcus epidermidis* [1, 2]. In addition, the anti-adhesive properties of Ta-based materials have been analyzed [3].

Due to their structural stability and biocompatibility, non-degradable polymer materials have been applied in a wide range of medical devices, in particular, as bearing surfaces in hip implants, vascular grafts or catheters and ocular implants [4, 5].

Hernioplastic mesh prostheses are commonly used for substitution of aponeurotic and muscle tissue. Polypropylene is probably the most widely used prosthetic material in mesh repair owing to its low cost, non-biodegradability and excellent incorporation [6]. The main properties required for intraperitoneal meshes are mechanical stability, adhesion prevention, good incorporation to prevent recurrence, shrinkage and mesh migration, as well as lack of inflammation and allergic reactions. The currently used polypropylene meshes induce severe chronic inflammatory reactions over a long-term period [7]. These processes lead to scarring and have a negative effect on the connective tissue and abdominal walls. Moreover, effects have been reported on the apoptosis rates due to the polymer destruction and their degradation products, resulting in carcinogenic potential based on apoptosis rates. In addition, the inflammatory reaction can lead to adhesions or fistula formation [8]. There is strong association between adhesions and bowel obstruction, infertility and chronic pain [9]. Obviously, further research is needed to improve the mechanical and anti-adhesive properties of modern meshes. It is known that the biocompatibility of polymer surfaces might be improved by surface modification. Several problems are associated with the magnetron discharge deposition of coatings on plastic substrates. First, the plastic melts easily, so that to avoid this process the coatings should be deposited at a low power of the magnetron discharge. Secondly, internal stresses in the films lead to deformation of the coatings due to the softness of the plastic materials. Under some deposition conditions, delamination is observed as a result of the very high internal stresses in the films.

The research presented was aimed at studying the effect of deposition process parameters on the structural and mechanical properties of Ta, Ta_2O_5 and TaON films on polypropylene meshes intended for hernioplasty application.

2. Materials and methods

Polypropylene plates and meshes were used as substrates (Fig. 1). A wide range of pressures were applied to optimize the deposition process and the optimal technological conditions were determined allowing one to avoid plastic and coatings deformation during the deposition process.

The samples were cleaned in an ultrasonic bath following a standard technique. Then, an ion source (Radical type) was used to clean the sample surface before deposition [10]. The deposition of Ta, Ta_2O_5 and TaON coatings by magnetron sputtering was performed in a high-vacuum pumping system with a base pressure of about 1×10^{-3}Pa. A tantalum target with a diameter of 170 mm was used. The magnetron discharge power was 4 – 5 kW. Also, an ICP source for oxygen activation with RF power of up to 1 kW was applied. The ion cleaning process parameters were $p_{Ar} = 1.2\times10^{-2}$ Pa, oxygen mass flow rate $q = 45$ sccm. The main deposition process parameters were $p_{Ar} = 1.2\times10^{-2}$ Pa, magnetron voltage $U_m = 465 - 480$ V, magnetron current $I_m = 2.3 - 3.4$ A, Ta_2O_5 deposition oxygen flow $q_O = 20$ sccm, TaON deposition nitrogen flow $q_N = 16$ sccm, oxygen flow $q_O = 4$ sccm.

The coatings thickness and adhesion properties were evaluated by standard techniques. The values measured of the coatings thickness were 1.1 μm (Ta), 1.05 μm (Ta_2O_5) and 0.8 μm (TaON). The surface morphology and topography were observed by a JSM-7100F electron scanning microscope (SEM, JEOL, Japan). The elemental distribution and chemical composition of the coatings were analyzed by energy dispersive X-ray (EDX) spectroscopy (Oxford Link ISIS 300). X-ray photoelectron spectroscopy (XPS) measurements were carried out using an ESCALAB MkII apparatus (VG Scientific, UK). The mechanical tests of the Ta-based coated samples to evaluate the Vickers hardness parameters and elastic modulus were performed on a G200 Nano Indenter (KLA, USA).

3. Results and discussion

Figure 2 shows SEM images of Ta, Ta_2O_5 and TaON coatings deposited by magnetron sputtering. The coatings cross-section was smooth without delamination. The EDX spectra revealed the presence of the main characteristic elements tantalum (Ta), oxygen (O) and nitrogen (N).

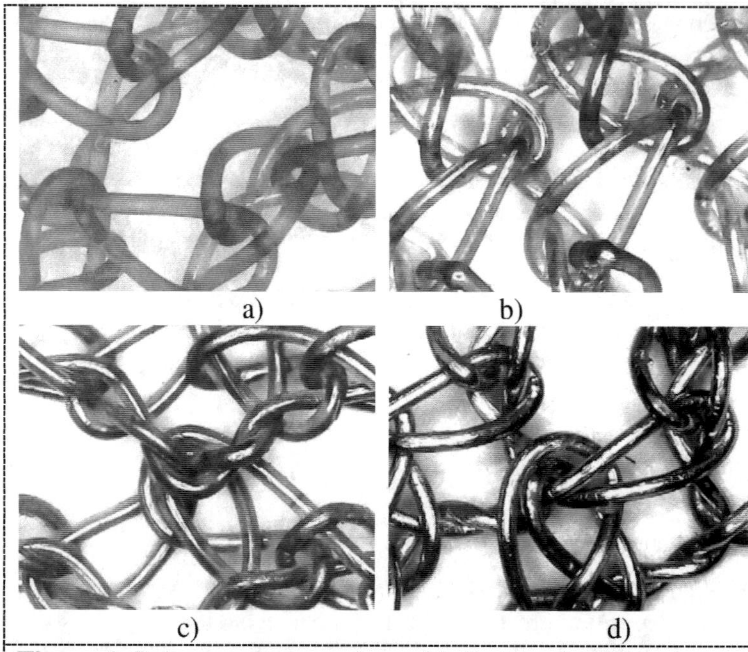

Figure 1. Polypropylene meshes with Ta-based coatings: a) uncoated, b) Ta_2O_5, c) TaON d) Ta

The compositional analysis of the Ta, oxide Ta_2O_5 and oxynitride TaON coatings by means of XPS revealed the photoelectron spectra of Ta4f, O1s, N1s coatings as previously observed [10]. The spectrum included the photoelectron lines of Ta (4f7/2) and Ta (4f5/2). The Ta^{+5} signals were detected at the binding energies of 26.8 eV and 28.7 eV. The O1s high-resolution spectrum exhibited a peak at the binding energy position of 530.9 eV associated with the Ta-O chemical bond. The N1s peak was detected at the binding energy of 396.2 eV assigned to the Ta-N chemical bond. This peak is generally considered to be evidence of O atoms replacement by N atoms in the Ta_2O_5 crystal lattice [11]. Also, a weak N1s peak at the binding energy of 398.0 eV was observed corresponding to the Ta-N-O chemical bonds [12].

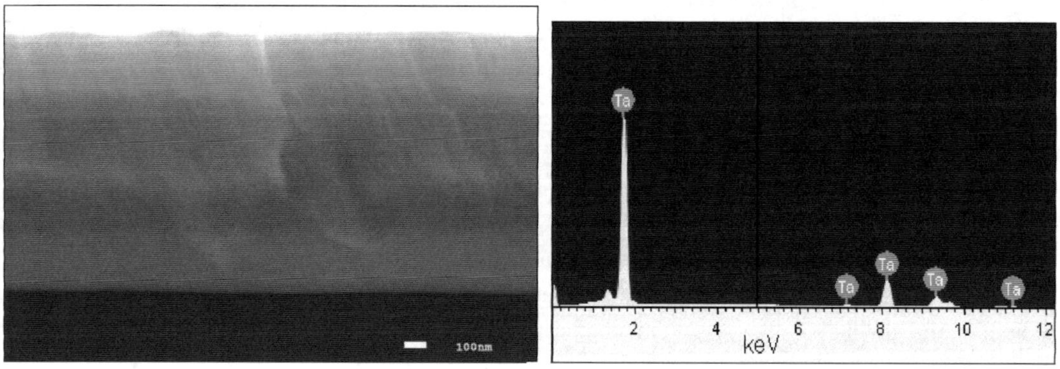

a

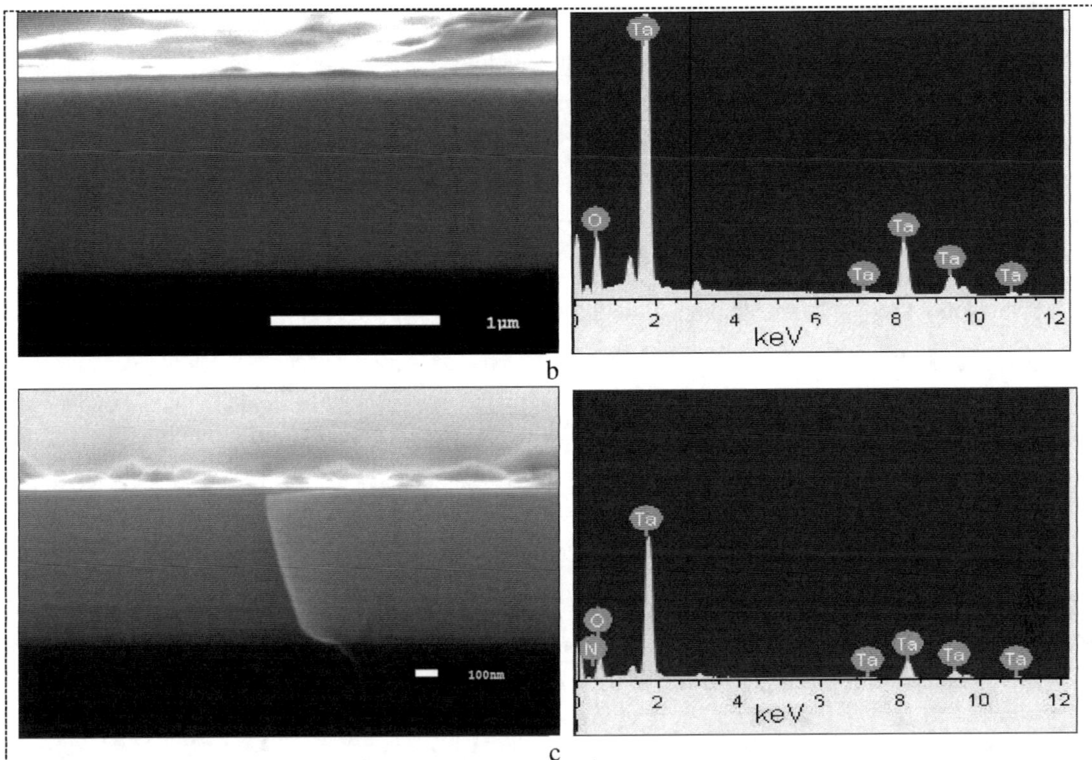

Figure 2. SEM images and EDX spectra of Ta (a), Ta_2O_5 (b), and TaON (c) deposited coatings.

The coatings' mechanical properties are strongly correlated with their microstructures, phase and chemical composition. The values of the hardness (*H*) and Young's modulus (*E*) of the Ta-based coatings compared with the polypropylene substrates are presented in Fig. 3.

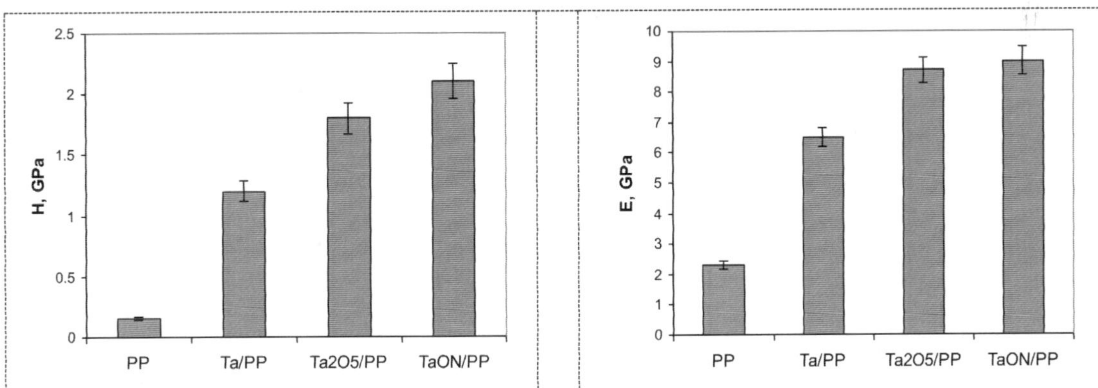

Figure 3. Hardness (*H*) and Young's modulus (*E*) of the Ta-based coatings compared with the polypropylene substrates.

The improvement of the coated materials' properties in comparison with the uncoated polymers was analyzed previously [7, 13]. The Ta-based coatings demonstrated a relatively high hardness and Young's modulus, corresponding to a high elastic recovery and a high resistance to cracking [14], which makes these materials highly attractive for hernioplasty applications. The variations in hardness and Young's modulus under irradiation processes were previously analyzed [15]. Both the hardness and the Young's modulus increased within micrometers-depth in the irradiated samples.

4. Conclusion

Ta-based coatings deposited by magnetron sputtering were found to improve the mechanical characteristics and biocompatibility of the polypropylene materials. The results demonstrated the effect of Ta-based coatings composition on their properties. The coated samples exhibited higher nano hardness and elasticity values, significantly exceeding the values of untreated polypropylene. Thus, optimal deposition parameter were achieved. The Ta-based films possessed properties favorable in view of potential prosthetic applications in hernioplastic surgery.

Acknowledgment

The authors acknowledge the financial support of the Bulgarian National Science Fund of the Ministry of Education and Science, Contract No KP 06 M 37/1 /06.12.2019, the scientific cooperation program of the Bulgarian – -Ukrainian Academies of Sciences, and the international scientific project No 8-03-20 of the National Academy of Sciences of Ukraine.

References

[1] Schildhauer T A, Robie B, Muhr G and Koller M 2006 *J. Orthop. Trauma* **20** 476
[2] Harrison P L, Harrison T, Stockley I and Smith T J 2017 *J. Bone Joint* **99B** 1153
[3] Zykova A, Safonov V, Yakovin S, Dudin S, Melnikova G, Petrovskaya A, Tolstaya T, Kuznetsova T, Chizhik S A and Donkov N 2020 *J. Phys.: Conf. Series* **1492** 012038
[4] Wu W, Cheng R, das Neves J, Tang G, Xiao J, Ni Q, Pan G, Li D, Cui W and Sarmento B 2017 *J. Controlled Release* **261** 318
[5] Minati L, Migliaresi C, Lunelli L, Viero G, Serra M D and Speranza G 2017 *J. Biophys. Chem.* 229: 151
[6] Costello CR, Bachman S L, Ramshaw B J, and Grant S A 2007 *J Biomed Mater Res B Appl Biomater* **83** 44
[7] Scheidbacha H, Tannapfelb A, Schmidta U, Lipperta H and Köckerling F 2004 *J. Eur Surg Res* **36** 313
[8] Lehle K, Lohn S, Reinerth G G, Schubert T, Preuner J G and Birnbaum D E 2004 Biomaterials **25** 5457
[9] Ellis H. 2005 *J Am Coll Surg* **200** 641
[10] Walkowicz J, Zavaleyev V, Zykova A, Safonov V, Dudin S, Yakovin S and Donkov N 2018 *J. Phys.: Conf. Series* **992** 012034
[11] Chun W J, Ishikawa A, Fujisawa H, Takata T, Kondo J N, Hara M, Kawai M, Matsumoto Y and Domen K 2003 *J. Phys. Chem. B* **107 (8)** 1798
[12] Matizamhuka W R, Sigalas I and Herrmann M 2008 *Ceram. Int.* **34 (6)** 1481
[13] Elinson V M, Shchur P A, Deshevaya E A and Kravets L I 2019 *J. Phys.: Conf. Series* **1313** 012016
[14] Leyland A and Mathews A 2004 *Surf. Coat. Technol.* **177/178** 317
[15] Brun C, Delobelle P, Fromm M, Berger F, Chambaudet A and Jaffiol F 2001 Mat. Sci. Eng. **A315** 63

The digital axiograph – a novel tool in bruxism prevention

I Taneva[1], T Uzunov[1], N Apostolov[1] and N Filev[2]

[1]Department of Prosthetic Dental Medicine, Faculty of Dental Medicine, Medical University-Sofia, 1 Georgi Sofiyski Str, 1431 Sofia, Bulgaria
[2]Chief Dental Technician, 141 Slivnitsa Blvd, 1233 Sofia, Bulgaria

e-mail: i.taneva@fdm.mu-sofia.bg

Abstract. Digital (electronic) axiography is an applicable promising technology with proven accuracy that allows registering the spatial location of the mandible and all lower jaw movements: mouth opening and closing, protrusion, left and right laterotrusion, dynamic function. In spite of bruxism treatment still being a controversial topic, occlusal splint application is proven to be the most efficient way of prevention. The purpose of the study is to develop a protocol of planning and producing preventive occlusal splints for bruxism management by means of digital axiography. The implementation of digital technologies reveals new possibilities in diagnosing, treating and tracking bruxism. With great attention to the individual trajectories of the lower jaw movements, the effectiveness of the occlusal splint prophylaxis is no longer hypothetical.

1. Introduction

Bruxism as a parafunction may result in abnormal tooth wear, mobility, fracture, intrusion, opening of contacts, drifting, erosion, or pulp pathology. Amongst the effects of bruxism on the dentition are pathologic tooth migration, bone alterations, temporomandibular joint disorders (TMD) and pain [1].

The use of occlusal splints for the management of the bruxing patient has been advocated for many years. Due to the contradictory nature of the parafunction, there is still no universal conclusion and its etiology is considered multifactorial [2]. Therefore no specific, reliable treatment capable of cancelling bruxism is yet available and all efforts are directed towards the prevention of bruxism's destructive effects [1]. When planned and produced precisely, occlusal splints in their variety successfully prevent the negative consequences of the parafunction [3].

The axiograph (also known as facebow) records the path of a condylar point or the path of a point in the vicinity of the condyle. Opening, closing, and protrusive movements have been investigated using axiography [4]. The clinical use of axiography includes location of the transverse horizontal axis and the detailed gathering of data needed to adjust a fully adjustable articulator [5].

Digital (electronic) axiography is an applicable promising technology with proven accuracy, equivalent to conventional axiographs. Due to the modern-day technical and software enhancements, it is superior to them in terms of data acquisition and interpretation [6]. Electronic axiography studies allow one to analyze such parameters as quality indicators, quantity indicators (range), symmetry, synchronicity of movement rates (between left and right temporomandibular joints, TMJ) [7].

A number of studies have investigated the diagnostic potential of digital axiographs in relation to programming an articulator according to the individual parameters of the patient's mandibular movements and its application in the diagnosis of TMJ [8]. There is a variety of systems available: Dentograf (Prosystom, Russia), Zebris (Amann Girrbach AG, Austria), Proaxis (Prosystom, Russia),

Cadiax (Whip Mix Corp., USA), Modjaw (Modjaw, France), Sicat JMT+ (Sicat GmbH, Germany). Despite the technological differences, digital functional diagnostics and jaw motion registration are integrated in modern-day dentistry [9].

2. Material and methods

2.1. Material

A 42-year old male was clinically diagnosed with sleep bruxism using a standard questionnaire and an oral examination. A BiteStrip® personal disposable miniature electromyography device was used to objectively confirm the diagnosis. The patient's complaints included visible signs of tooth attrition, stiffness in the masseter muscles area and occasional clicking in the right temporomandibular joint (TMJ). The patient had all teeth of the dentition present (third molars included) and no prosthetic restorations. To exclude any particular temporomandibular disorder (TMD) and for the aims of the digital axiography, cone-beam computerized tomography (CBCT) was administered.

Registration and analysis of the position and the movement trajectories of the mandible were performed by a Dentograph (Prosystom, Russia) digital (electronic) axiograph. A novel device for complex functional diagnostics of the lower jaw articulation, it consists of the following parts (figure 4):

- a three dimensional (3D) camera, embedded in a special head piece;
- a central marker and two side markers equipped with a grid of sensors designed to be detected by the camera on the head of the patient;
- a software program (P-Art, Prosystom, Russia) developed particularly to process the data obtained by the axiograph.

The patient's upper and lower jaw, intercuspation (occlusal relation) included, were scanned by a powderless intraoral scanner Medit I500 (Medit Corp., South Korea, Figure 1). The scanner has an optimized workflow and interface (Figure 2 and 3) and works efficiently at high speed. It provides a high level of accuracy, which was crucial for the purpose of this study's scans of the upper and lower dental arch.

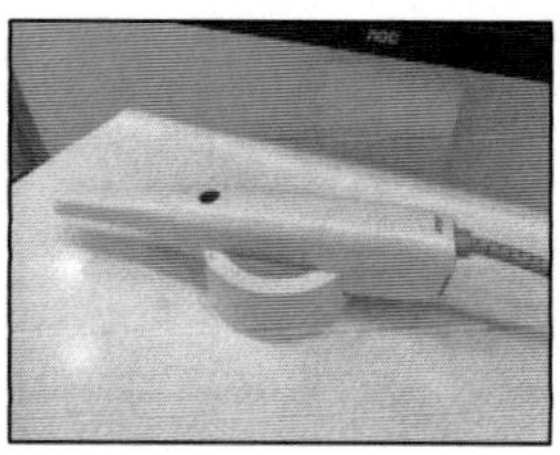

Figure 1. Medit I500 intraoral scanner.

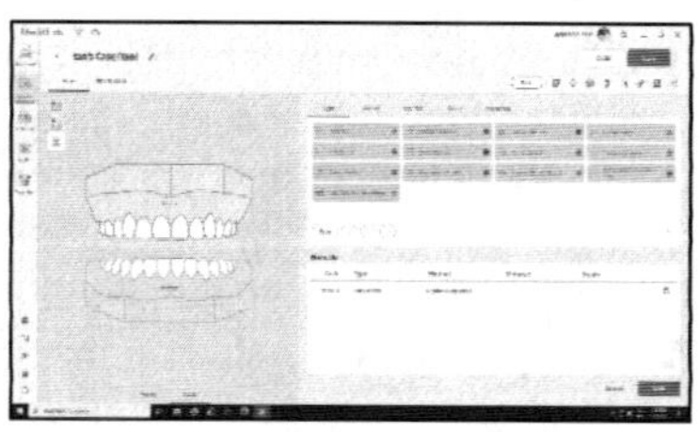

Figure 2. Medit I500 interface.

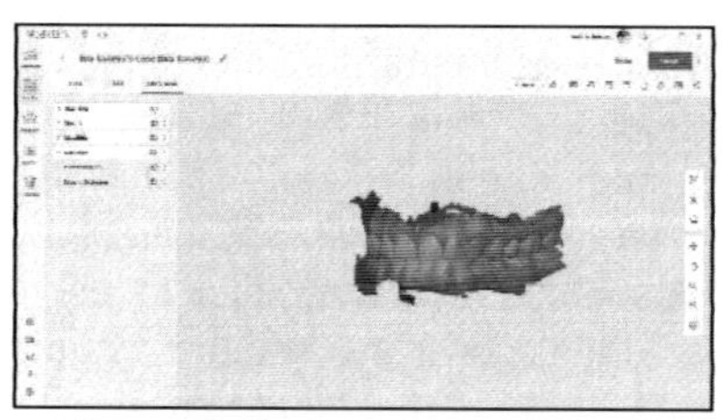

Figure 3. Scanned upper and lower dental arch.

The design of the occlusal splint was carried out in Exocad (Exocad GmbH, Germany) – dental software for computer-aided design (CAD). The splint was manufactured by means of 3D printing [10] with Form 2 (Formlabs Inc., USA) digital stereolithography (SLA) printer (Figure 5 (a), with the material selected being Dental LT Clear liquid photopolymer resin developed for particularly for occlusal splint production. For the mandatory post polymerization processing of the material, the attendant Formlabs devices were used (Form Wash – Figure 5 (b) and Form Cure – Figure 5 (c).

IOP Conf. Series: Materials Science and Engineering 1056 (2021) 012012 doi:10.1088/1757-899X/1056/1/012012

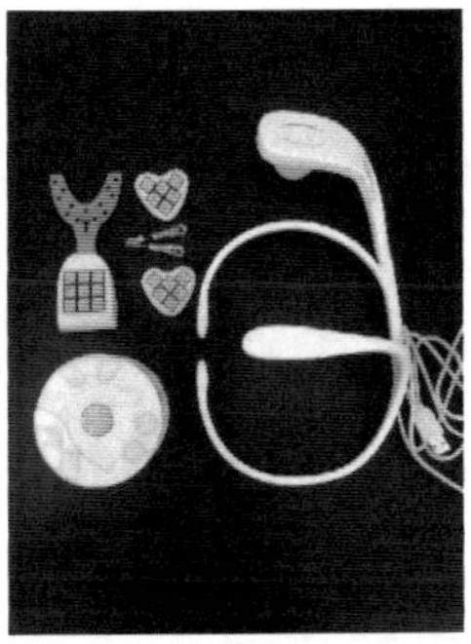

Figure 4. Digital axiograph Dentograf – full setup.

Figure 5. Form 2 SLA printer (a), Form Wash ultrasonic bath (b) and Form Cure ultraviolet polymerization device (c).

2.2. Methods

To fulfil the aim of this study and based on the instructions given by the manufacturer, the following steps of performing a digital axiography were formulated:

1) Registration with the central marker.

Bite registration additive silicone material is applied onto the axiograph's bite fork (Figure 6) and with regard to its indentation for obtaining the sagittal plane the fork is fixed to the upper dental arch. The patient closes the mandible until contact with the fork to support it (Figure 7). Preferably, this can be executed by means of two dental cotton rolls. Once the silicone sets, the central marker is switched on and the spatial position of the prosthetic plane is registered through wireless connection (Figure 8 (a) and (b). Registration is visible in the software simultaneously.

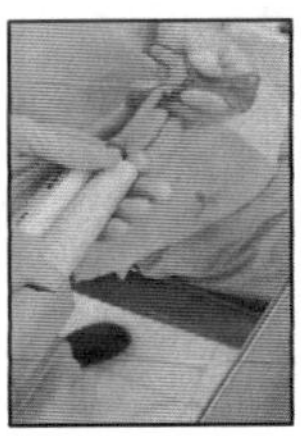

Figure 6. Bite registration

Figure 7. Central marker fixed to the upper dental arch.

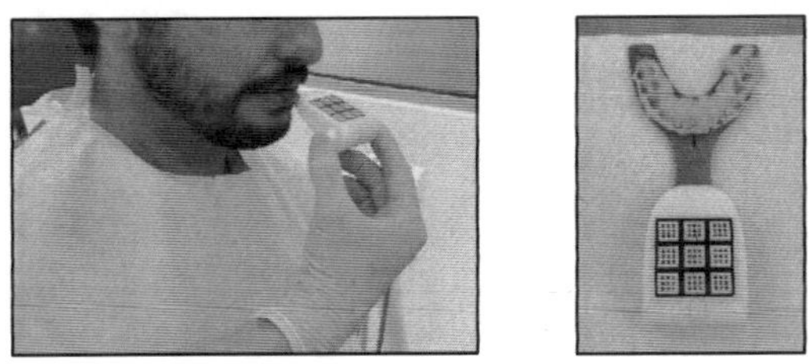

Figure 8. Central marker switched on – (a) and after performing the registration – (b).

2) Fixing the side markers (Figure 9):
   a. Marker on the maxilla – situated in I. quadrant (right), canine area;
   b. Marker on the mandible – situated in III. quadrant (left), canine area.

Fixation of the side markers can be achieved by using a photopolymer composite (paste consistency) or glass ionomer cement (for fillings). It is recommendable to have a dental assistant retract the lips of the patient until the fixing material used is set. It is important to follow the direction sign in the outer corner of the sensor grid on each side marker – the sign should be distally placed; this gives the proper left and right position (Figure 10).

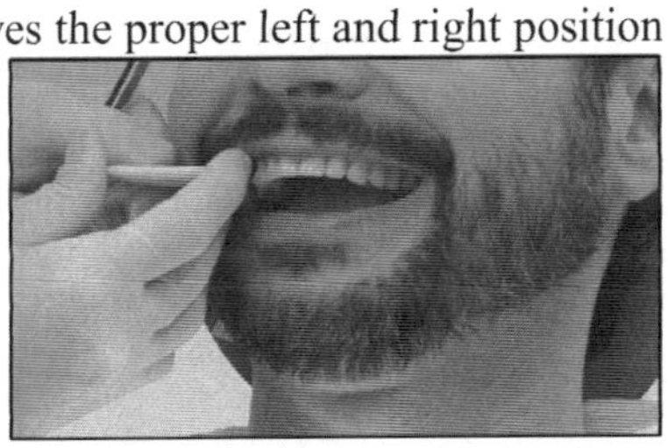

Figure 9. Fixing the side markers.

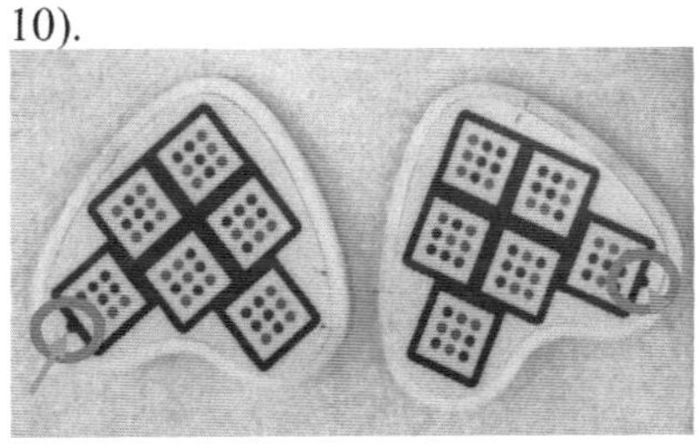

Figure 10. Direction sign on the markers.

IOP Conf. Series: Materials Science and Engineering 1056 (2021) 012012 doi:10.1088/1757-899X/1056/1/012012

3) Placing the head piece (3D camera) on the head of the patient and connecting it with the computer.
4) Taking reference frames:
   a. First reference frame – central marker and right (maxillary) marker (Figure 11);
   b. Second reference frame – right and left (mandibular) marker (Figure 12).

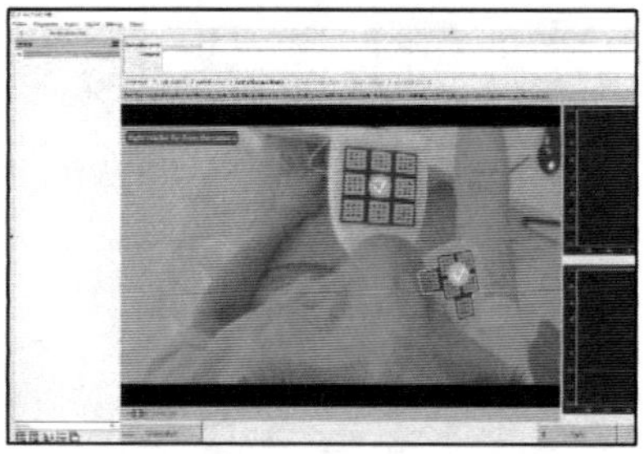

Figure 11. Taking first reference frame.

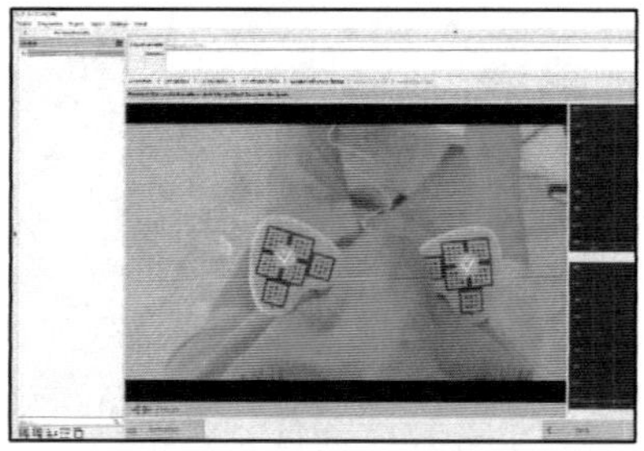

Figure 12. Taking second reference frame.

5) Axiography record.

Recording the spatial location of the mandible and all lower jaw movements: mouth opening and closing, protrusion, left and right laterotrusion, dynamic function (can also be achieved by the patient chewing gum). While being recorded, all trajectories are visible in different color in the software.

6) Analysis of the results.

This final step is executed in the software entirely. We chose to merge all data in the Complex diagnostics module of the program:

- Upload of individual models of the patient – intraoral scans of maxilla, mandible and occlusion. Occlusal contacts are also made visible;
- Upload of CBCT scans of the patient – the CBCT findings are united with the models from the intraoral scan according to dentition markers; nevertheless, the scans alone can be used in the software for the creation of virtual models.
- Alignment of the individual models – using an intraoral scan with the bite fork returned in the patient's mouth. Part of the fork should be also visible in the scan.

7) Export to CAD software (Exocad).

After running the analysis, the final step is to export the individual models of the patient in current position, the recorded trajectories of the lower jaw (lower incisal, right and left articular), values of all lower jaw angles (sagittal, frontal, transversal) and the alignment of the upper jaw individual model on the bite fork. By completing the import with these parameters, the need of using a virtual articulator in Exocad is optional.

8) Occlusal splint design in Exocad.
9) Export of the design in Form 2 SLA printer's software, material selection, printing and post processing.

3. Results

The methodology described in subsection 2.2. is approbated and proves fully applicable for functional diagnostics of the position, movement and trajectories of the lower jaw. The validated protocol steps are clearly formulated and the objective of each one is well defined. This protocol has been applied as a complex solution to bruxism cases and proves to be superior and more precise than the methods using average parameters.

4. Discussion

The implementation of digital technologies revealed new possibilities in diagnosing, treating and tracking bruxism and any accompanying TMJ dysfunctions. The higher accuracy of the method used was ensured by the supplementary tomography data.

The axiograph's software offers the options of importing scanned individual plaster models (semi-digital method) or of calculating individual parameters for mounting plaster models in a fully adjustable mechanical articulator (regular analog method).

The semi-digital method allow one to avoid the us of an intraoral scanner – impressions are taken from patient's upper and lower jaw with additive silicone impression material, plaster models are poured from the impressions and scanned in a desktop optical scanner (Ceramill Map 600, Amann Girrbach AG, Austria). In cases where it is necessary to work with a virtual articulator, the virtual models from any of the two types of scans can be uploaded and mounted. The movement trajectories can also be visualized and the splint is designed in the CAD software.

In the analog method, the plaster models are mounted in an Artex CR (Amann Girrbach AG, Austria) fully-adjustable articulator. For this purpose, the alignment is used of the bite fork on the special 3D stand preliminarily adjusted according to the values calculated in the software. While the bite fork with the bite registration material from the patient is on the 3D stand, the upper jaw plaster model is placed on the bite fork. Then, the mechanical articulator is adjusted to the angle values given from the axiography. The bite splint can be manufactured from wax directly on the models and then polymerized by means of lost wax technique and polymethyl methacrylate (PMMA) heat curing resin.

If comparing the methods in terms of the number of laboratory steps, the analog method is the most time consuming. At greatest risk of error is again the transfer of the upper jaw's spatial position giving the prosthetic plane in the mechanical articulator. On the other hand, when the same transfer is done in a virtual articulator, there is no potential risk of error occurrence. If comparing the approbated protocol with the semi-digital approach, the intraoral scans performed are the most accurate. A number of mistakes can be made when taking regular impressions: from mixing and placing the impression material in the impression tray, to its temperature, to an improper impression technique of the dentist. Furthermore, when casting the impressions the expansion of the plaster must correspond to the shrinkage of the impression material used. Even with the most precise desktop optical scanner one can possibly scan any mistake just as precisely. Any inaccuracy it the process of planning, producing and adjusting the occlusal splint can result in inducing bruxism.

5. Conclusion

The developed thorough digital workflow algorithm for treatment of the bruxing patient allows a complete and predictable outcome. With every single individual parameter of the patient considered, the occlusal splint manufactured is in full accordance with the end purpose – prevention of the parafunction. Digital axiography provides the opportunity of repeating the recordings under the exact same conditions, which makes it a proficient method for patient follow up. With great attention to the individual trajectories of the lower jaw movements, the effectiveness of the occlusal splint prophylaxis is no longer hypothetical.

Contemplating the lower jaw's dynamic behavior has intrigued researchers in the field for many years. With digital axiography, appropriate treatment in accordance with every patient's individual features is effectively attainable.

Acknowledgements
The research was funded by the National Scientific Program "Young scientists and post-doctoral students 2020".

References

[1] Paesani D A 2010 *Bruxism: Theory and Practice* (Berlin: Quintessenz Verlags-GmbH) pp 3-60
[2] Bader G Lavigne G J 2000 *Sleep Med. Rev.* vol 4 pp 27-43

[3] Luchowski L Tomaka A A Skabek K Tarnawski M and Kowalski P 2016 *Information Technologies in Medicine (ITIB 2016 Advances in Intelligent Systems and Computing)* ed E Pietka *et al* (Springer) vol 472 pp 267-73
[4] Porter R Poyser N Briggs P and Kelleher M 2007 *Dent. Update* vol 34 pp 198-207
[5] Slavicek R 1988 *J. Clin. Orthod.* vol 22 pp 656-67
[6] Kucukkeles N Ozkan H Ari A and Çilingirtürk A 2005 *J. Prosthet. Dent.* vol 94 pp 190-4
[7] Wagner A Seemann R Schicho K Ewers R and Piehslinger E 2003 *J. Prosthet. Dent.* vol 90 pp 503–9
[8] Antonik M M Kalinin Iu 2011 *Stomatologiia (Mosk)* vol 90 pp 23-7
[9] Tamaki K Čělar A G Beyrer S and Aoki H 1997 *J. Prosthet. Dent.* vol 78 pp 373–78
Bakalczuk M Bozyk A and Iwanek M 2004 *Ann. Univ. Mariae Curie-Sklodowska Med.* vol 59 pp 324-7
[10] Franklin P, McLelland and Brunton P 2010 *Eur. J. Prosthodont. Restor. Dent.* vol 18 pp 17-22
[11] Chkhikvadze T V Bekreev V V Roshchin E M Trufanov V D Yurkevich R I and Ivanov S Yu 2019 *Sovrem. Tekhnologii Med.* vol 11 pp 111
[12] Taneva I and Uzunov T 2020 *J. Phys.: Conf. Ser.* vol 1492 pp 012018

Optically-controlled density of alkali atom vapor for biomagnetic sensing applications

S Tsvetkov[1] **and S Gateva**[1]

[1]Institute of Electronics, Bulgarian Academy of Sciences, 72 Tsarigradsko Chaussee blvd., Sofia 1784, Bulgaria

stocvet@ie.bas.bg

Abstract. Biomagnetic sensing is a powerful non-invasive tool for sensing the extremely small magnetic signals of living organisms. In recent years, the efforts to achieve higher resolution has brought about a demand for novel methods of increasing this tool's sensitivity and further its miniaturization. In this paper, we present our studies on the optical control of the atomic vapor density in antirelaxation coated cells for application in optical magnetometry. A special system for homogeneous illumination of the cell walls is used to increase the atomic density, avoid the deposition of alkali atoms on the cell walls and reduce the coating deterioration.

1. Introduction

Biomagnetic sensing is a powerful non-invasive method for registration of the magnetic fields induced by the currents in living organisms, such as magnetocardiogram (MCG), which detects the magnetic field from the heart, magnetomyogram (MMG), from the muscles, and magnetoencephalogram (MEG), from the brain. The main concept behind biomagnetism has to do with ions that move between the living cells and induce extremely small magnetic fields compared to the magnetic interference from the surrounding environment. In order to detect such small biomagnetic signals, very sensitive magnetometers are needed [1,2].

Since the optical (atomic) magnetometers (OM) reached the sensitivity of the superconducting quantum interference devices (SQUID), optical magnetometry has been evolving very rapidly and finding more and more applications. An OM does not require cryogenic cooling and can be positioned closer to the object of interest; also, it does not necessitate shielding during the measurements [3,4]. In an OM, a pumping light polarizes the spin of the atoms. As this polarization is sensitive to magnetic fields, it changes in an external magnetic field B. The change of the spin orientation is detected optically by a probing laser light and a polarization detector.

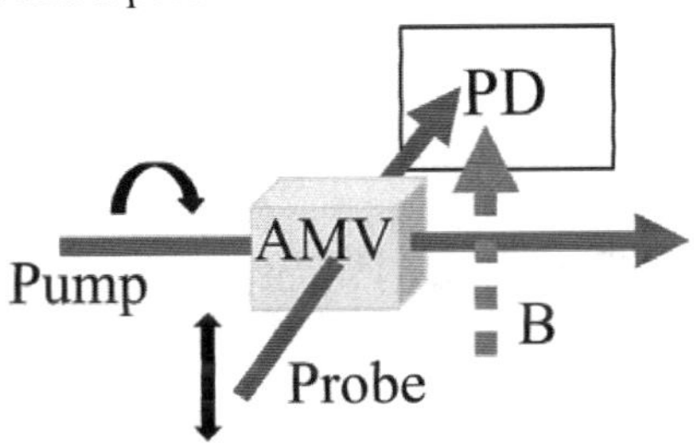

Figure 1. AMV – alkali metal vapor, B – magnetic field, PD – polarization detector.

As an example, the OM described in [5] is shown schematically in figure 1. A circularly polarized pump laser beam orients the atomic spins. The orientation of the atomic spin is sensitive to the magnetic field *B*. This change is sensed by a linearly polarized probe beam.

The uncertainty of the magnetic field measurement performed for a time *T* with an ensemble of *N* atoms with coherence time τ is

$$\delta B = \frac{1}{g\,\mu_B}\frac{\hbar}{\sqrt{N\tau T}}, \quad (1)$$

where μ_B is the Bohr magneton, g is the ground-state Landé factor [4].

The sensitivity of the measurements increases with the square root of *N*, *T* and τ. A way of enhancing the sensitivity of the measurements is to increase the spin-relaxation time τ, which is basically limited by the depolarization of the atomic spins when the atoms collide with the walls of the vapor cell. In order to increase the depolarization time, a buffer gas is usually employed; or, alternatively, an anti-relaxation (AR) coating, such as paraffin, octadecyltrichlorsilane (OTS), polydimethylsiloxane (PDMS), is applied on the inner surface of the optical cells [6,7]. High-quality coatings allow up to 10^6 bounces without relaxation. The number of atoms increases with the temperature. However, high temperatures reduce the coatings' AR properties, because the alkali metal atoms adsorbed on the coating degrade the AR coatings and consequently shorten the lifetime of the optical cell [8]. As a result of the necessity for high density and good AR properties in a very small optical cell, the temperature chosen should not be very high, while the coating itself should remain stable at this temperature.

In general, optical magnetometers make use of alkali atoms. The most appropriate elements from the alkali metals are rubidium (Rb), cesium (Cs), and potassium (K), because their density is sufficiently high even at room temperature.

An alternative way to increase the density of alkali metal atoms confined in an optical coated cell is the light-induced atomic desorption (LIAD). It is a non-thermal process that allows one to control the alkali metal vapor density in AR coated cells by illuminating the optical cell walls with non-coherent and non-resonant light [9]. Upon illumination, the alkali atoms that have been previously adsorbed onto the surface of the coating are desorbed so that the vapor density increases. When the light is turned off, the atoms are adsorbed back on the coating or pumped by the stem, or accumulate on some part of the surface of the coating forming atomic clusters or a monolayer as a result of the inhomogeneous illumination [7,10].

In this communication, we report our investigations on controlling the optical atomic density in optical cells with possible applications in optical magnetometry.

2. Experimental setup

The experimental setup is shown in figure 2. A special system for homogeneous illumination of the cell walls is used to increase the atomic density, to avoid the deposition of alkali atoms on the cell walls, and to reduce the deterioration of the coating or accumulation of atoms [11]. It consists of a sphere coated inside with a high-reflection layer that reflects the desorbing light homogeneously over a coated optical cell; a high-power 460-nm LED; and a controller maintaining the LED current thus controlling the illumination intensity, which is compared to the intensity measured by a feedback photodiode inside the sphere.

A 6-cm long vacuum optical cell is used filled with a natural mixture of Rb isotopes. The inner walls of the cell are coated with a PDMS AR coating, which acts as a Rb dispenser when the desorbing light is switched on. The cell has an open stem filled with solid or liquid Rb depending on the temperature. Without illumination (LIAD), the density in the volume of the cell is defined by the temperature of the coldest point of the cell. In the case of cells with AR coatings, the density usually is slightly lower than that defined by the temperature due to the absorption of atoms by the coating.

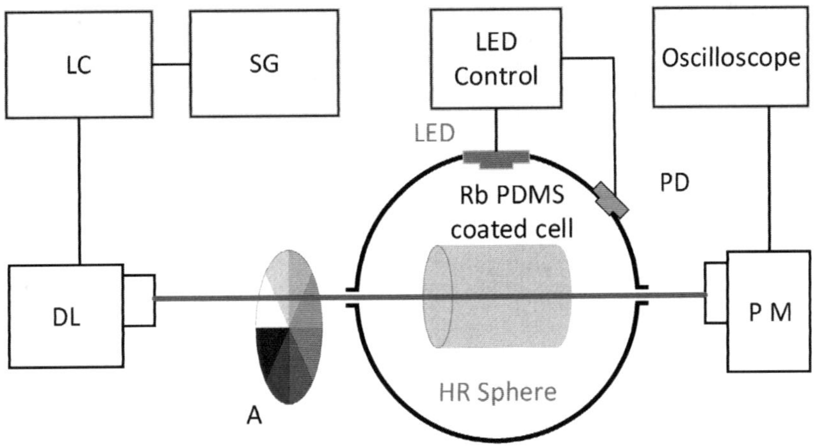

Figure 2. Experimental setup
(DL – 780-nm diode laser, HR Sphere – a high-reflectivity sphere, PM – power meter, A – attenuator, LC – laser current and temperature controller, SG – signal generator, PD – photodiode).

3. Results and discussion

The Rb atoms density in the cell is estimated from the absorption of the 780-nm diode laser tuned in resonance with the D2 Rb line. The optical pumping, which is very strong in AR coated cells [12], is made negligible by reducing the laser light power to 4 μW. The absorption coefficient $\alpha(\nu)$ of the medium can be written in terms of $\sigma(\nu)$, the atomic absorption cross-section and N, the number density of the atomic vapor

$$\alpha(\nu) = N\sigma(\nu). \quad (2)$$

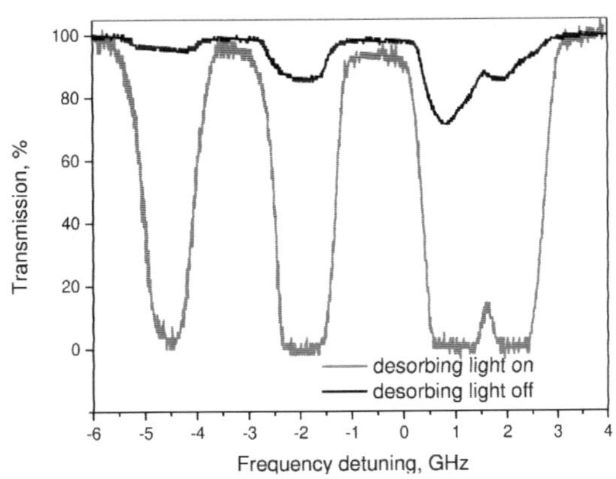

Figure 3. Transmission of the Rb D2 line in a PDMS coated cell with (red curve) and without (black curve) LIAD.

In figure 3, the Rb D2 line transmission spectra in PDMS coated cell are compared for two cases: when the desorbing light is not switched on and when light with intensity of 5.5 mW/cm^2 illuminates homogeneously the whole cell. As seen, three of the four Doppler-broadened profiles of the Rb D2 line are completely saturated and the vapor is optically thick; however, no reduction of the absorption due to photon-reabsorption [13] is observed. The calculated atomic density in this case is $N = 2\times10^{11}$ cm^{-3}, which is an increase of almost two orders of magnitude compared with the initial density.

LIAD in the case of homogeneous illumination not only significantly increases the atomic density at room temperature, but also removes atoms from the surface, thus restoring the initial AR properties of the coating. Therefore, the LIAD might contribute to increasing the sensitivity in two ways: increasing the density without decreasing the AR properties of the coating, and by removing the layers or clusters of metals on the surface of the AR coating and in this way increasing the relaxation time.

The results of previous experimental works [8,14], based on different methods of measuring the number of collisions without depolarization of the atomic spin, have shown that the relaxation rate when the density is controlled by LIAD is lower than the relaxation rate when the density is controlled via the temperature.

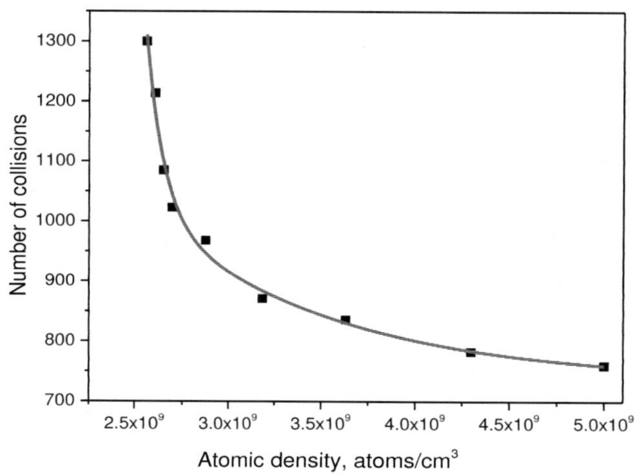

Figure 4. Number of collisions without spin relaxation when the atomic density is controlled by LIAD.

Figure 4 shows the dependence of the number of collisions without spin relaxation on the atomic density when LIAD is applied. After the initial sharp drop in the number of collisions without spin relaxation, a further increase in the atomic density by LIAD shows results in the curve flattening and approaching a constant value. This is not the case with a temperature density control, when the density continues decreasing steeply [14].

4. Conclusions

The investigations performed showed the potential of the optical control of the atomic density for the future development of magnetic sensors. The use of LIAD results in better AR properties at lower temperatures, which will increase the sensitivity of the OM sensors as a result of the decreased alkali atoms relaxation rate. The lower operating temperature of these sensors will extend their life as higher temperatures degrade the AR properties of the coatings over time. Lower operation temperatures will also aleviate the need for thermal insulation and consequently will favor the detection of weaker magnetographic signals. All this will benefit the future development of new magnetic sensors with higher sensitivity and longer lifetime.

Acknowledgments
This work was supported by the National Science Fund of Bulgaria (Contract No DN 08-19 /2016 "New coherent and cooperative effects in hot alkali vapor". S.T. acknowledges the National Program "Post-Doctoral students" RMS No 271/2019.

References

[1] Krause A H-J and Dong H 2017 Biomagnetic Sensing, in *Label-Free Biosensing: Advanced Materials, Devices and Applications*, eds M J Schoening and A Poghossian, Springer Series on Chemical Sensors and Biosensors, Springer International Publishing AG 449-474
[2] Paek AY, Kilicarslan A, Korenko B, Gerginov V, Knappe S and Contreras-Vidal JL 2020 Towards a Portable Magnetoencephalography Based Brain Computer Interface with Optically-Pumped Magnetometers in *42nd Annual International Conference of the IEEE Engineering in Medicine & Biology Society (EMBC)*, Montreal, QC, Canada 3420-3423
[3] Kitching J 2018 *Appl. Phys. Rev.* **5** 0313021
[4] Budker D and Romalis M 2007 *Nature Physics* **3** 227–234
[5] Knappe S, Sander T and Trahms L 2014 Optically-pumped magnetometers for MEG. In: *Magnetoencephalography*, Springer, Berlin, Heidelberg
[6] Chi H, Quan W, Zhang J, Zhao L and Fang J 2020 *Appl. Surf. Sci.* **501** 143897
[7] Li W, Balabas M, Peng X, Pustelny S, Wickenbrock A, Guo H and Budker D 2017 *J. Appl. Phys.* **121** 063104
[8] Graf M, Kimball D, Rochester S, Kerner K, Wong C, Budker D, Alexandrov EB and Balabas MV and Yashchuk VV 2005 *Phys. Rev. A* **72** 023401
[9] Alexandrov E, Balabas M, Budker D, English D, Kimball D, Li C and Yashchuk V 2002 *Phys. Rev. A* **66** 042903
[10] Yi Y, Robinson H, Knappe S Maclennan JE, Jones CD, Zhu C, Clark NA and Kitching J 2008 *J. Appl. Phys.* **104** 023534
[11] Tsvetkov S, Taslakov M and Gateva S 2017 *Appl. Phys. B* 123:922017
[12] Siddons P, Adams C, Ge C and Hughes I 2008 *J. Phys. B: At. Mol. Opt. Phys.* **41** 155004
[13] Matsko A , Novikova I, Scully M and Welch G 2001 *Phys. Rev. Lett.* **87** 13
[14] Tsvetkov S, Gateva S and Cartaleva S 2019 *AIP Conf. Proc.* **2075** 030010

Atmospheric particulate matter pollution over residential urban areas during COVID-19 quarantine

R Ilieva[1], B Angelova[1], M Iliev.[1], D Stoyanov[2], V Groudeva[1], Z Cherkezova-Zheleva[3] and I Nedkov[2]

[1] - Sofia University "St. Kliment Ohridski", Faculty of Biology, Bulgaria
[2] - Bulgarian Academy of Science, Institute of Electronics, Bulgaria
[3] - Bulgarian Academy of Science, Institute of Catalysis, Bulgaria
E-mail: nedkovivan@yahoo.co.uk

Abstract. This study of particulate matter (PM) pollution was conducted in May and June 2020 during the COVID-19 lockdown in residential areas of the Bulgarian capital Sofia. Its methodology was based on lidar monitoring of the ground layer of the atmosphere and focused on the sampling procedure. The sampling was performed with portable instruments *in situ* in the areas with air pollution fixed on the city's map during the lidar monitoring. The collected PM was examined by a set of physical methods (XRD, SEM, EDAX) and PCR tests for viral contamination). A wide range of data on the nature of PM were obtained. Data on the number of COVID-19 positive individuals in the town for the period was also summarized.
Key words: particulate matter (PM), bioaerosol, lidar monitoring, COVID19 lockdown period.

1. Introduction

The subject of this study is the particulate matter (PM) carried by the atmospheric bioaerosol. The term PM is collective expression for particles and usually describes the solid-state part of water droplets [1, 2]. In assessing the PM pollution, it is widely accepted to work with the mass concentration of PM 2.5 μm and PM 10 μm. In nature, PM have different origin, their size varies widely and their morphology creates the conditions for them to be carriers of various inorganic and bioorganic pollutants. Moreover, their water envelope in the atmospheric aerosol can become a breeding ground for bacteria, fungi and viruses, so that they can present danger to human health. On 11 March 2020, the World Health Organization (WHO) declared the coronavirus pandemic. The new virus Sars-Cov-2 (COVID-19) is poorly studied and spreads very rapidly among the human population. It is mainly transmitted by inhalation of aerosol droplets. The conventional view [3] of viral spread uses the term "short-range transmission" and assumes that droplets ejected by the infected subject are heavy and spread over short distances. However, when this process takes place outdoors, the droplets become part of the aerosol environment and could undergo a different way of spreading. The diameter of these diverse particles is normally very small, they sediment slowly and are easily conveyed by air currents in the so-called "long distance transmission" [2]. This was confirmed by a study in 2010, when Chen and al. [4] reported that the avian flu virus was found in large concentrations in the air after dust storms. European authors reported recently on observing a relationship between the spread of the Sars-Cov-2 virus and air pollution [5]. The present work offers an extended study of the atmospheric bioaerosol during a COVID 19 lockdown over densely populated urban areas of the capital of Bulgaria, the city of Sofia, and provides extensive information on the type of PM (size and morphology) and their inorganic and organic content, including bacteria and fungi.

IOP Conf. Series: Materials Science and Engineering 1056 (2021) 012014 doi:10.1088/1757-899X/1056/1/012014

2. Methods and devices

The lidar mapping was performed by a lidar system installed in the Lidar Station of the Institute of Electronics, Bulgarian Academy of Sciences (IE-BAS). The laser emitter (wavelength of 510.6 nm) is a pulsed CuBr vapor laser with a repetition rate of 10 kHz at a 15-ns pulse duration. The PM sampling was implemented using an Andersen Cascade Impactor. Nutrient agar with cycloheximide was used for bacteria and YGC agar for fungal enumeration. The positive hole method was used for total colony count expressed as CFU/m^3. The control of the particles by size and mass concentration in this experiment was expanded through the systematic use of a Microcontroller SDS 011 – PM2.5 and PM10 during the lidar experiments and in the daily experimental practice. Additionally, the PM was collected on # 3μm filters during the lidar monitoring. The collected PM was studied by SEM, EDAX, Mössbauer spectroscopy and PCR tests for viral contamination.

3. Results and discussion

This study was conducted in May and June 2020, when a COVID-19 lockdown was strictly observed. The subject of the study were the residential areas (Mladost and lower part of Lozenets districts - see Fig.1a) of Sofia. The formal control of PM urban air pollution is focused mainly on the content of PM2.5 and PM10. Three stationary licensed stations of the Executive Agency (EA) at the Ministry of Environment and Water are located in the areas studied and provide daily online information on the PM concentration.

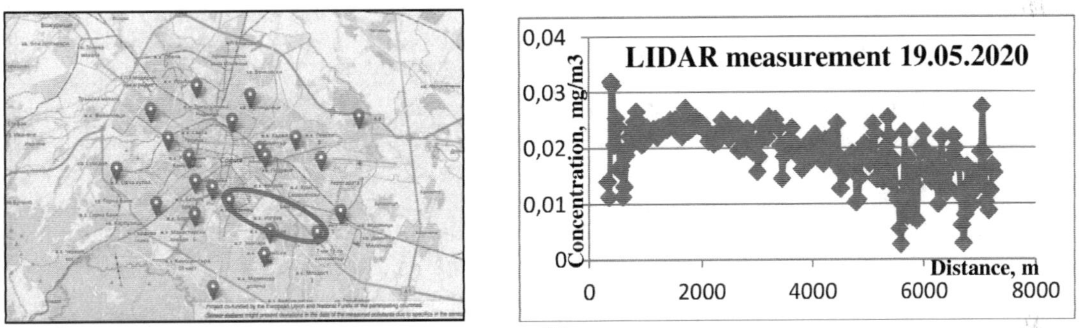

a) b)

Figure 1. a) Map of the city Sofia showing the fixed licensed stations and outlining the area under study; **b)** Data on the variation of the mass concentration of PM during the month of May in an area 7 km away from the Lidar Station

The designated area was subject to lidar monitoring. The characteristics of the lidar scan adopted in this case are discussed in detail in our chapter of the book [6]. Below we will briefly describe the basic expressions and the lidar data processing. The two major lidar parameters, the extinction coefficient $\alpha(r)$ and the backscatter coefficient $\beta(r)$, are calibrated in terms of the aerosol mass concentration following the well-known method in [7] and making use of the mass concentration M_a data obtained by the sampling device. For the lidar ratio $LiR=\alpha(r)/\beta(r)$ we adopted the typical value of $LiR=50$ [7]. The parameters $\beta(r)$ and $\alpha(r)$ were calculated using the lidar equation under the assumption of a horizontally-homogeneous atmosphere:

$$P(r)=P_0\frac{c\tau}{2}C\frac{\beta(r)}{r^2}exp\left[-2\int_{r_0}^{r}\alpha(r)dr\right] \quad (1)$$

where $P(r)$ is the power of the detected laser radiation backscattered from the atmosphere from a distance $r = \frac{ct}{2}$ after a period of time t following the moment of laser pulse emission, and τ is the pulse duration. The calibration dependencies were constructed of $\alpha(r)$ and $\beta(r)$ on the mass concentration in $[mg/m^3]$. In both cases, the linear fit (y = A + Bx) shows acceptable values of the standard deviation (less than 4 %) and the correlation coefficient (over 0.92). The plots can be used directly for calibrating the lidar maps. More detailed information about the lidar monitoring during the spring period of 2020 can be found in our work [8]. Fig. 1b illustrates a typical example of the remote sensing capabilities.The dependence of the PM concentration calculated on the basis of the extinction in the aerosol of the surface atmosphere at a height of up to 30 m is presented for a distance of 7 km from the Lidar Station, with the whole scanned city area being about 24.4 km^2. The mass concentration varies in a small range from 0.01 to about 0.03 mg/m^3. The maximums of PM pollution in the urban atmosphere are within the permissible norms for the whole period of studies.

The quality of the lidar sensing deteriorates in rainy weather and strong solar radiation and does not allow direct differential analysis in terms of particle size. During the experiment, this reduced the number of lidar experiments and necessitated the development of the current methodology, which expends the possibilities for objective daily assesment. An important element of the methodology is sampling. For this purpose, in areas with increased air pollution fixed on the city plan by the lidar monitoring, *in situ* sampling was performed using the following portable instruments: 1. Absorber with filters that collect PM $\geq$ 1 μm in size. For this purpose, a portable device with 4 filter sockets is used. The device reads the water content and the solid part of the aerosol; 2. Cascade Impactor, which separates the PM by size and allows one to study their microbial content; 3. Electronic sensors, which are calibrated with respect to data of the licensed stations [8] and give information about the mass concentration of PM2.5 and PM10.

After sampling, the material collected from devices 1 and 2 was examined by a set of physical methods (XRD, SEM, EDAX, MöS) and PCR viral contamination tests. This enabled us to obtain a wide range of data on the PM crystallo-chemical composition and morphology, on the nature of the surface and on their microbiota and elemental composition. The studies showed a wide variety of particles, which strongly depended on the seasonal characteristics, such as flowering and humidity. A common feature is that mostly aggregates of particles are observed, where in a peculiar way they combine inorganic and bioorganic material. Fig. 2 summarizes the weekly data for May (Fig. 2a) and June (Fig. 2b) from the measurements of PM2.5, PM10, bacterial contamination and the presence of fungi. Tables 1 and 2 show the meteorological data for the months considered. The X-ray diffractograms recorded show that the examined PM are amorphous – no crystalline phases were detected. The MöS analyses of the PM show the presence of iron ions in the second valence of oxidation, mostly in the atmosphere of Lozenets district, while equal amounts of divalent and trivalent iron were registered in the samples from Mladost district. More information was presented in our work [8]. Quantitative analyses were performed to determine the total number of cultured microorganisms as follows: aerobic heterotrophic bacteria (NA medium) and fungi (YGC medium). The samples were collected by means of the six-stage Cascade Impactor and performed in triplicate for each sample.

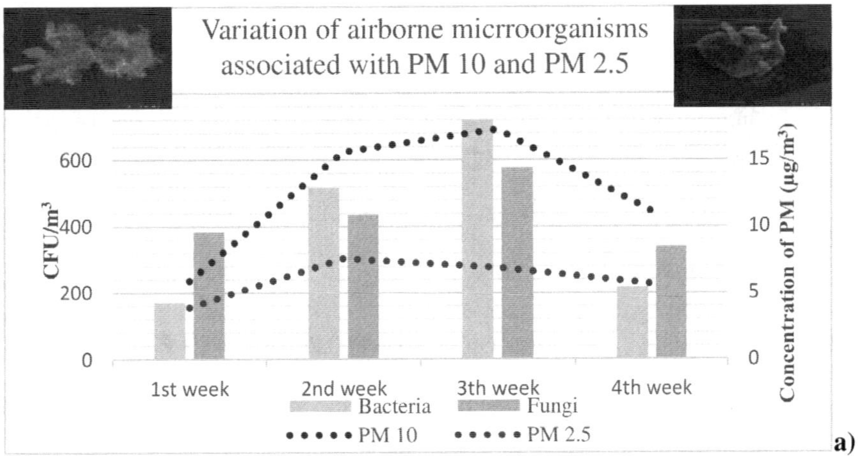

Table 1. Meteorological data for May 2020

Meteo Data	1st week	2nd week	3rd week	4th week
T **(°C)**	12 ± 2	19 ± 3	27.3 ± 1.8	18 ± 3.1
RH (%)	46.9 ± 3	45.3 ± 12	41.5 ± 3.2	64.5 ± 3.5
Wind (m/s)	5.65 ± 0.65	1.9 ± 0.7	1.7 ± 0.2	3.5 ± 5

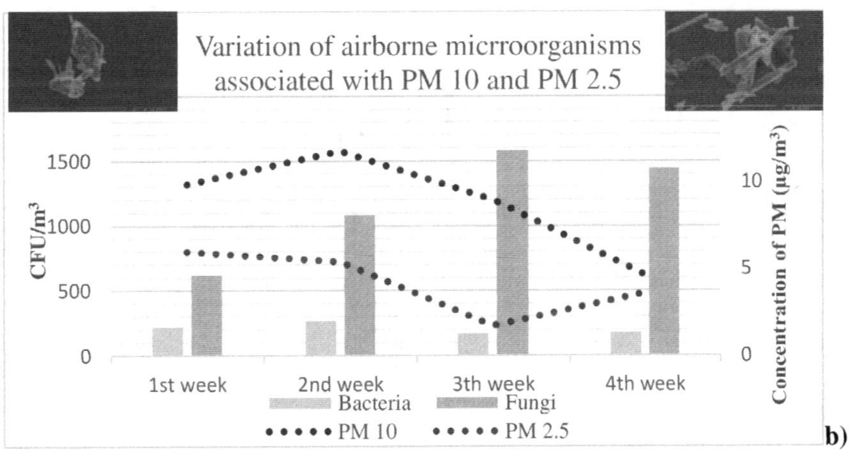

Table 2. Meteorological data for June 2020

Meteo Data	1st week	2nd week	3rd week	4th week
T **(°C)**	24.5 ± 1.5	20.25 ± 3.75	24 ± 4	21.6 ± 4.5
RH (%)	44.5 ± 14.5	65.7 ± 15.2	61.7 ± 13	70.8 ± 17
Wind (m/s)	1.8 ± 2	3.57 ± 4.2	4.6 ± 4	1.7 ± 1

Figure 2. Monthly variation in the mass concentration of PM2.5, PM10, bacteria and fungi contents **a)** in May and Table 1 and **b)** in June and Table 2 with weekly meteorological-data. Insets – SEM images of PM.

The results of the analysis show a significantly higher number of bacteria and fungi in both time intervals of the study. When comparing the data from this analysis over time: May, 2020 and June, 2020, a significant increase in the number of the two studied groups is reported. As already mentioned, the first interval tested in May is characterized by extremely low anthropogenic pressure due to the lockdown sanitary-epidemiological restrictions for the period (Fig. 2a). Typical for the period shapes of PM are presented in the insets. The figures show that the pollution with PM is within the standard permissible limits, while in the first two weeks of June an increase of PM2.5 and PM10 is seen. The bacterial contamination in the two months is within 200 CFU/m^3, the content of fungi in the air increases sharply in June up to 1600 CFU/m^3.

An attempt was made to compare these data with the official data of national institutions monitoring the epidemic. Based on these data, Fig.3 shows the number of people infected with COVID-19 in the city of Sofia. The analyses of the data on PM and microbial contamination in the context of COVID-19 lockdown at first glance shows an increase in the infected individuals occurring 3-4 days after the observed increase in PM2.5, PM10 and fungi in the area. Nevertheless, we do not undertake to make a definitive relationship assessment, unlike our colleagues in works [4, 5]. The main problem comes from the fact that in a large city like Sofia with more than 1.3 million inhabitants it is incorrect to draw conclusions on the number of infected in the whole city using data pertaining to areas with a population of 160 thousand. In addition, during the experiment, the air pollution was below the generally accepted norms and should not have posed a danger.

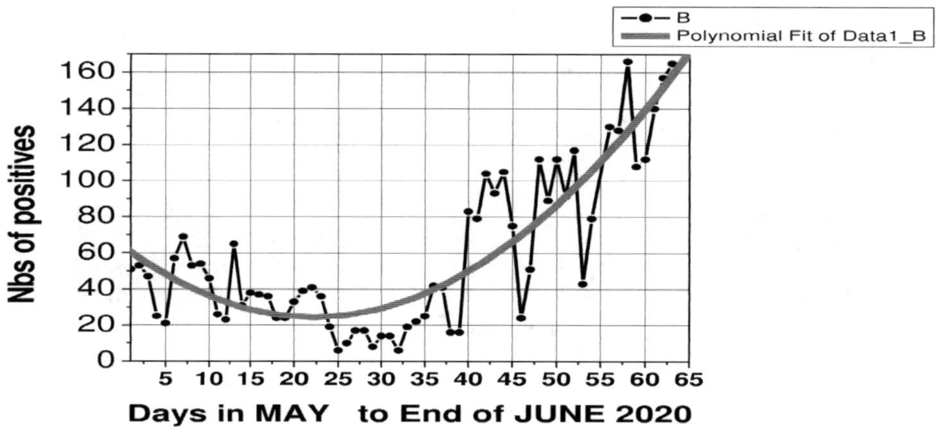

Figure 3. Daily change in the number of COVID-19 positive persons during the period discussed (May-June 2020)

4. Conclusions

Combining lidar monitoring with *in situ* sampling and examination of the solid part of the PM is the main goal of this work. During the experiment, the air pollution did not exceed the generally accepted permissible limits. The strong dependence of the microbial pollution on the seasonal features is impressive. The increase in temperature and humidity in June greatly increases the presence of fungi in the air. The area studied is residential with many parks with tall trees than emit pollen during the period, while last year's rotting organic matter on the ground causes an increase in the fungi content. The search for correlation between air pollution and viral epidemics is not the main task of this work; however, in June an increase of the COVID-19 positive cases was registered 4-5 days after an increase of PM2.5 and PM10. These results give us grounds to continue these studies, as they open prospects for obtaining in-depth results on the relationship between the PM content and the COVID-19 epidemic, thus creating preconditions for rapid prevention.

Acknowledgments

This work was financed by contract DN18/26 with the National Science Fund, Bulgaria, and included in the European Program of the COST Action CA16202.

References

[1] Environmental monitoring of bioaerosols at regulated facilities, Environment Agency, July 2018

[2] Fuzzi S., Baltensperger U., Carslaw K., Decesari S, Denier van der Gon H.,Facchini M.C., Fowler D., Koren I., Langford B., Lohmann U., Nemitz E., Pandis S.,Riipinen I., Rudich Y., Schaap M., Slowik J.G., Spracklen D.V., Vignati E., Wild M.,Williams M.M., and Gilardoni S, 2015. Particulate matter, air quality and climate: lessons learned and future needs. *Atmos. Chem. Phys.*:**15**:8217–8299. DOI: 10.5194/acp-15-8217-2015

[3] Macher J. 1999. (Ed.) Bioaerosols: Assessment and Control, *American Conference of Governmental Industrial Hygienists*, Cincinnati; 322 p. DOI: ISBN: 978-1-882417-29-1

[4] Chen P.S., Tsai F.T., Lin C.K., Yang C.Y., Chan C.C., Young C.Y., Lee C.H. 2010. Ambient influenza and avian influenza virus during dust storm days and background days. *Environ. Health Perspect.* **118**:1211–1216. doi: 10.1289/ehp.0901782

[5] Comunian S., Dongo D., Milani C., Palestini P., 2020. Air Pollution and COVID-19: The Role of Particulate Matter in the Spread and Increase of COVID-19's Morbidity and Mortality, *Int J Environ Res Public Health.* **17**(12):4487. Published online 2020 Jun 22. doi: 10.3390/ijerph17124487

[6] Stoyanov D., Nedkov I., Grudeva V., Cherkezova-Zheleva Z., Grigorov I., Kolarov G., Iliev M., Ilieva R., Paneva D., Ghelev C. 2019. Long-Distance LIDAR Mapping Schematic for Fast Monitoring of Bioaerosol Pollution over Large City Areas. *Atmospheric Air Pollution and Monitoring.* Ed. A. Lakhouit, ISBN: 978-1-78985-280-6, Publ. IntechOpen, London (2019)

[7] Weitkamp C. 2005. (Ed.) Lidar Range-Resolved Optical Remote Sensing of the Atmosphere, *Springer Series in Optical Sciences*, Springer. 456 p. DOI:https://doi.org/10.1007/b106786

[8] Angelova B., Iliev M., Ilieva R., Grigorov I., Kolarov G., Paneva D., Kolev Ch., Z. Cherkezova-Zheleva, Groudeva V., Stoyanov D., Nedkov I. 2020. Complex study of bioaerosol composition of atmosphere over urban areas based on lidar monitoring during the lockdown COVID-19, *Machines. Tecnologies. Materials*, Issue **7**/2020, pp.272-276

AUTHOR INDEX